LOW-VOLTAGE CMOS OPERATIONAL AMPLIFIERS

Theory, Design and Implementation

THE KLUWER INTERNATIONAL SERIES IN ENGINEERING AND COMPUTER SCIENCE

ANALOG CIRCUITS AND SIGNAL PROCESSING

Consulting Editor
Mohammed Ismail
Ohio State University

Related Titles:

ANALYSIS AND SYNTHESIS OF MOS TRANSLINEAR CIRCUITS, Remco J. Wiegerink
ISBN: 0-7923-9390-2
COMPUTER-AIDED DESIGN OF ANALOG CIRCUITS AND SYSTEMS, L. Richard Carley, Ronald S. Gyurcsik
ISBN: 0-7923-9351-1
HIGH-PERFORMANCE CMOS CONTINUOUS-TIME FILTERS, José Silva-Martínez, Michiel Steyaert, Willy Sansen
ISBN: 0-7923-9339-2
SYMBOLIC ANALYSIS OF ANALOG CIRCUITS: Techniques and Applications, Lawrence P. Huelsman, Georges G. E. Gielen
ISBN: 0-7923-9324-4
DESIGN OF LOW-VOLTAGE BIPOLAR OPERATIONAL AMPLIFIERS, M. Jeroen Fonderie, Johan H. Huijsing
ISBN: 0-7923-9317-1
STATISTICAL MODELING FOR COMPUTER-AIDED DESIGN OF MOS VLSI CIRCUITS, Christopher Michael, Mohammed Ismail
ISBN: 0-7923-9299-X
SELECTIVE LINEAR-PHASE SWITCHED-CAPACITOR AND DIGITAL FILTERS, Hussein Baher
ISBN: 0-7923-9298-1
ANALOG CMOS FILTERS FOR VERY HIGH FREQUENCIES, Bram Nauta
ISBN: 0-7923-9272-8
ANALOG VLSI NEURAL NETWORKS, Yoshiyasu Takefuji
ISBN: 0-7923-9273-6
ANALOG VLSI IMPLEMENTATION OF NEURAL NETWORKS, Carver A. Mead, Mohammed Ismail
ISBN: 0-7923-9049-7
AN INTRODUCTION TO ANALOG VLSI DESIGN AUTOMATION, Mohammed Ismail, José Franca
ISBN: 0-7923-9071-7
INTRODUCTION TO THE DESIGN OF TRANSCONDUCTOR-CAPACITOR FILTERS, Jaime Kardontchik
ISBN: 0-7923-9195-0
VLSI DESIGN OF NEURAL NETWORKS, Ulrich Ramacher, Ulrich Ruckert
ISBN: 0-7923-9127-6
LOW-NOISE WIDE-BAND AMPLIFIERS IN BIPOLAR AND CMOS TECHNOLOGIES, Z. Y. Chang, Willy Sansen
ISBN: 0-7923-9096-2
ANALOG INTEGRATED CIRCUITS FOR COMMUNICATIONS: Principles, Simulation and Design, Donald O. Pederson, Kartikeya Mayaram
ISBN: 0-7923-9089-X
SYMBOLIC ANALYSIS FOR AUTOMATED DESIGN OF ANALOG INTEGRATED CIRCUITS, Georges Gielen, Willy Sansen
ISBN: 0-7923-9161-6

LOW-VOLTAGE CMOS OPERATIONAL AMPLIFIERS

Theory, Design and Implementation

by

Satoshi Sakurai
National Semiconductor

Mohammed Ismail
Ohio State University

KLUWER ACADEMIC PUBLISHERS
Boston / Dordrecht / London

Distributors for North America:
Kluwer Academic Publishers
101 Philip Drive
Assinippi Park
Norwell, Massachusetts 02061 USA

Distributors for all other countries:
Kluwer Academic Publishers Group
Distribution Centre
Post Office Box 322
3300 AH Dordrecht, THE NETHERLANDS

Library of Congress Cataloging-in-Publication Data

A C.I.P. Catalogue record for this book is available
from the Library of Congress.

Printed on acid-free paper.

Printed in the United States of America

Contents

List of Figures

List of Tables

PREFACE

In this book, the theory, design and implementation of low-voltage (<3V) CMOS operational amplifiers are discussed. Both single- and two-stage architectures are treated. Opamps with constant-g_m input stage are designed and their excellent performance over the rail-to-rail input common mode range is demonstrated. The work presented here is a result of Ph.D dissertation research conducted by Satoshi Sakurai at the Ohio State University, with Professor Mohammed Ismail as his adviser. The project was initiated in the Spring of 1992, after the first set of CMOS constant-g_m input stages were introduced by a group from Technische Universiteit Delft and Universiteit Twente, The Netherlands. These earlier versions of circuits are discussed in this book along with new circuits developed at the Ohio State University. The design, fabrication (MOSIS Tiny Chips), and characterization of the new circuits were completed in June, 1994.

Readers are presumed to have some understanding of basic analog integrated circuit design concepts in order to fully appreciate the work presented. However, the topics are presented in a logical order and the circuits are explained in great detail, such that the book can be read and enjoyed by those without much experience in analog circuit design.

In the first part of the book, motivations behind the work are stated. The necessity for the reduction in the power supply voltage is discussed in Chapter I, and the advantages of having rail-to-rail input stage with constant-g_m characteristics are pointed out.

In Chapters III, IV, and V, constant-g_m input stages are presented and their operations are explained. The new input stages introduced in Chapter V are used in the design of various opamps. The design of these opamps and their computer simulation results are given in Chapters VI, VII, and VIII.

Performances of opamps fabricated using MOSIS service are presented in Chapter IX where the effectiveness and the usefulness of the constant-g_m input stages are clearly demonstrated. Measurement techniques used, *e.g.* for measuring the transconductance, the gainbandwidth, the phase margin etc., are described together with process parameters used and SPICE netlists of all circuits.

We wish to thank all those who assisted us, the Semiconductor Research Corporation for funding this work, and our families for their support and understanding.

Satoshi Sakurai, *Santa Clara, CA*
Mohammed Ismail, *Columbus, Ohio*

Chapter 1

Introduction

In recent years, low voltage VLSI circuits have received considerable attention in the field of microelectronics. One is bound nowadays to come across a technical report titled "A 3.3-V CMOS ***(some kind of a digital circuit or a system)", when reading a VLSI related journal[1]. The trend in the lower power supply voltage is a natural byproduct created by the advancement in the technology of CMOS transistor scaling.

1.1 Background

The constant reduction of the minimum feature size of a MOS transistor has been ongoing for the past few decades[2, 3]. The main reason has been to integrate more and more transistor circuits into a single silicon chip, and in particular, to put more *digital* circuits on a chip, and reduce the cost of products utilizing such ICs. Since digital circuits are designed using minimum-feature-size transistors, shrinking the transistor leads to reduction in the circuit area and cost. However, this is not always true in *analog* circuits.

The scaling of a transistor is 3-dimensional and hence the gate oxide thickness is reduced along with the channel length and width. As the channel length is scaled down into deep-submicron and the gate oxide becomes only a several nano-meters thick, the maximum allowable power supply voltage must be reduced to about 3V from the currently used 5V in order to ensure the device reliability[4, 5, 6].

While the role of digital circuits is increasing in present-day signal processing systems, the popularity of analog circuits is decreasing due to their design complexity, particularly that computer-aided analog design tools have not reached the maturity of their digital counterparts. However since we live in an analog world, we cannot avoid processing analog signals and in many cases it is more natural to use analog, rather than digital signal processing. The art of CMOS analog circuit design has been maturing in the last decade and the idea of integrating analog circuits on a digital chip followed naturally to further reduce cost. Such systems are called mixed-mode ICs[7]-[9]. Since the digital part of mixed mode ICs occupies most of the silicon real estate, the electrical parameters of MOS transistors are optimized for digital circuits. When analog circuits are designed using minimum-feature-size transistors, their performances generally degrade[10, 11]. However, since analog circuits consume only a fraction of the chip area, using larger device sizes to maintain good circuit performances does not increase chip size by very much. For example, if the analog part of a chip consumes 5% of the whole chip, increasing the analog part by 100% increases the chip size by only 5%.

One of the toughest challenges imposed on analog circuits in scaled CMOS processes is the reduction of the power supply voltage. It is nearly a two volt decrease in the maximum signal swing while the transistor threshold voltage, V_T, remains almost constant. V_T, which is about 1V could have easily been scaled down. However, in order to maintain a sufficient noise margin in digital circuits such as dynamic logics, V_T cannot be decreased much lower than 1V[12]. This fact puts severe restrictions on the architecture of analog circuits. For example, we cannot stack three diode connected MOS transistors and expect them to operate in the strong inversion region.

1.2 Significance of the Research

One of the most successful analog CMOS design techniques is the switched capacitor technique[13]-[17]. The time constant of a switched capacitor filter(SC-filter) is determined by the clock frequency and the ratio of capacitors which can be made very accurate. The major component of the SC-filter, the operational amplifier(opamp), often imposes limitations on both the small signal and large signal handling capabilities. Because SC-filters process the signal in a discrete-time manner, they need to be preceded by a continuous-time(CT) anti-aliasing filter and proceeded by a CT smoothing filter. These CT-filters may consist of opamps, and they can be implemented in CMOS technology using similar components as those used in SC-filters; such CT-filters are known as MOSFET-C filters[18]-[20]. It is evident that the opamp is a key player in analog signal processing. Opamps are used with negative feedback and assumed to be ideal during the design of opamp based circuits. However, it is well known that the finite gain-bandwidth product(GB) of opamps limits their use to lower frequency applications. In the past few years, operational transconductance amplifier-capacitor(OTA-C) filters or $g_m - C$ filters[21]-[24] have been implemented in MOS technology. Since the $g_m - C$ filter utilizes the transconductance, g_m, of the OTA to realize its time constant, the circuit can be operated at much higher frequencies compared to opamp based filters. However, because OTAs are used in an open loop manner, their linear input range directly affects the dynamic range of the entire filter. Also, the time constant determining parameter, g_m, is dependent on the MOS transistor parameters, thus the time constant is not as accurate as in opamp based SC-filters. OTAs are easily tuned by bias current or bias voltage to compensate for random parameter variations. OTAs with a large input range can be designed in CMOS technology and they are often called CMOS linearized voltage to current(V to I) converters or simply linear transconductors[25]-[36]. As different as the ways opamps and OTAs are used, their design goals and procedures are not the same. Thus, the effort in this research will be focused on opamps as they are no doubt still the most widely used analog building blocks.

With the reduction of power supply voltage to 3V, many of the existing CMOS analog building blocks, designed to operate from higher supply voltages, will lose a significant amount of operating range. Opamps will be affected of course due to this change, especially at the input stage. Thus, circuit design techniques which regain the operating range while maintaining the performance of the opamp as ideal as possible must be developed.

The significance of this research is that once circuit design techniques which enable rail-to-rail circuit operation without any performance deviation and degradation are developed, integration of analog and digital circuits on a CMOS process optimized for digital circuits becomes feasible for 3-V operations. Since CMOS technology is likely to remain as the dominant technology for VLSI, and since the 3-V power supply will be the standard for many years to come, success of this research should lead to development of future analog and mixed-signal VLSI design.

1.3 Organization of the Book

So far, it has been made clear that this research will focus ultimately on the design of CMOS operational amplifiers which can operate with 3V power supply voltage. The main focus will be to realize input stages which operate rail-to-rail, *i.e.*, with the entire power supply range, while maintaining the transconductance, g_m, constant.

In Chapter II, the effect of reduced power supply voltage on the performance of typical building blocks used in opamps is investigated. It will be shown that rail-to-rail signal swing at both the input and the output stages is necessary. Also, it will be shown that the simple rail-to-rail input stage realized by placing n- and p-channel differential pairs in parallel has a total input transconductance, g_{mT}, that is a strong function of the common mode input voltage, V_{CM}. Hence, the need for a constant-g_m input stage will be revealed and discussed.

In Chapter III, techniques used in previously developed CMOS constant-g_m input stages[37, 38] are reviewed and the problem, which might make the practical realization of these techniques difficult, is pointed out. This is the source of motivation for this research.

In Chapter IV, new techniques for realizing a constant-g_m input stage are introduced, and the complete input stages are introduced in Chapter V which contains the analysis of the input stage with second order effects of transistors taken into account.

In Chapter VI, rail-to-rail class AB output stage and the opamp architecture to be used in the proceeding chapters are introduced and studied.

In Chapters VII and VIII, a total of eight opamps are designed and computer-aided simulation results are given. In Chapter VII, the new constant-g_m input stages are used in the implementation of two single-stage opamps along with a simple rail-to-rail opamp without the constant-g_m characteristics. Two-stage versions of these opamps are given in Chapter VIII where the fully-differential versions of the two-stage opamps with constant-g_m input stage are given, also.

In Chapter IX, experimental results obtained from two MOSIS Tiny Chips are provided. These chips contain several opamps and test structures which make it possible to evaluate the constant-g_m input stage by itself, and also to characterize the MOS transistors.

Finally, conclusions and future directions are discussed in Chapter X.

Chapter 2

Operational Amplifiers in 3-V Supply

2.1 Introduction and Background

Before we actually start to discuss transistor circuits, some basic material and general information which can be referred to throughout this work are given.

Basic MOS Equations: For the most part of the analysis and design procedures, transistors are assumed to be operating in the saturation region. Transistors are biased such that this condition is satisfied under normal operation. The substrate terminal of every transistor is connected to the appropriate rails; thus, transistors are treated as three-terminal devices and the circuit schematics do not explicitly show the substrate connections. For an n-channel MOS transistor operating in saturation in the strong inversion region the drain current is given by

$$I = K(V_{GS} - V_{Tn})^2 \tag{2.1}$$

where V_{Tn} is the threshold voltage and $K = \frac{1}{2}\mu_n C_{ox}\frac{W}{L}$ is the transconductance parameter of a transistor. Unless otherwise specified, the transconductance parameter of a transistor M_i is to be given by K_i. Similarly the drain current for a p-channel transistor is given by replacing V_{Tn} and μ_n by V_{Tp} and μ_p in (2.1).

Many times in this book, the gate to source voltage of a transistor expressed as a function of the drain current will be used; this can easily be done by using an alternate form of (2.1) as follows:

$$V_{GS} = \sqrt{\frac{I}{K}} + V_T \tag{2.2}$$

The transconductance of a MOS transistor biased with a current I is given by

$$g_m = 2\sqrt{KI}; \tag{2.3}$$

this expression will be used throughout.

Simplified Opamp Model: A simplified opamp model which is composed of ideal circuit elements is shown in Fig. 2.1. There are two input terminals, V_{in1} and V_{in2}. The signals applied to these terminals are $V_{CM} + \frac{1}{2}v_{in}$ and $V_{CM} - \frac{1}{2}v_{in}$, respectively where V_{CM} is the common mode voltage and v_{in} is the input differential signal. The input signal $V_{in1} - V_{in2} = v_{in}$ is applied across R_{in}(considered to be infinity) and is converted to a current $g_{mT}v_{in}$ where g_{mT} is the total transconductance of the input stage. In the case of a single-stage opamp, the output voltage is V_{O1} and it is produced by sinking $g_{mT}v_{O1}$ into the resistor R_{O1}. For a two-stage design, the signal V_{O1} becomes the input signal to the second stage or the output stage. The transfer function of this two stage opamp model is given as follows.

$$\frac{v_{out}(s)}{v_{in}(s)} = \frac{g_{mT}g_{mO}R_{O1}R_{O2}(1 - sC_C/g_{mO})}{1 + as + bs^2}. \tag{2.4}$$

where

$$\begin{aligned} a &= R_{O1}(C_{O1} + C_C) + R_{O2}(C_{O2} + C_C) + g_{mO}R_{O1}R_{O2}C_C \\ b &= R_{O1}R_{O2}[C_{O1}C_{O2} + C_C(C_{O1} + C_{O2})]. \end{aligned}$$

Note that it may seem that V_{CM} does not affect anything since it is canceled at the input stage; however, when V_{CM} becomes close to one of the rails, the parameter g_{mT} is no longer a constant and becomes dependent on V_{CM} unless a circuit technique is developed to keep it constant.

Circuit Simulations: All the circuit simulations are performed using a CAD tool, APLAC[39]. The BSIM transistor model is used instead of commonly used LEVEL 2 model. Although the BSIM model has more accurate modeling for the transition between the strong and the weak inversion region than LEVEL 2 model, it still has some deviations from the actual transistor behavior, especially when it comes to simulating g_m of a MOS transistor. Both LEVEL 2 and the BSIM models have a drain current which is continuous between the strong and the weak inversion regions.

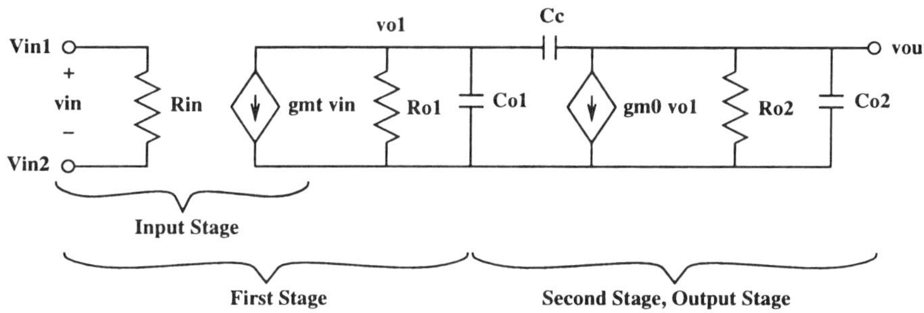

Figure 2.1: A simplified model of a two stage opamp.

However, only the BSIM model has a continuous transconductance, which is a derivative of the drain current with respect to the gate to source voltage, between these two regions. Moreover, neither BSIM nor LEVEL 2 model the moderate inversion region which exists between the two regions in an actual MOS transistor. The transconductance of an n-channel transistor simulated using both models are shown in Figure 2.2. The drain and the source terminals are connected to 3V and 0V, respectively, while the gate terminal is swept between 0.7V and 1.1V where V_T is approximately 0.9V. For the transistor simulated with LEVEL 2 model, we can clearly see the discontinuity in g_m when V_{gate} is about 0.9V. For the BSIM case, g_m is still not perfectly smooth, but it is much better than that obtained with LEVEL 2 model at least for the purpose of simulating g_m.

The BSIM parameters for a $2\mu m$ p-well process obtained from MOSIS are given in Appendix A along with the LEVEL 2 parameters of the same process. The circuit net-lists of every simulated circuit in this book is provided in Appendix B.

2.2 CMOS Building Blocks

Many CMOS opamps[40] evolved from their bipolar counterparts[41], and hence possess similar characteristics. However, since MOS transistors require larger turn on voltage and saturation voltage(for active region operation) than bipolar transistors, the signal swing of a CMOS circuit is usually more limited than that of the bipolar version for a given circuit topology. For the usual differential pair as an input stage of an opamp, with 1-V V_T and a 3-V power supply, the input common mode range is less than 2V. That is, more than 30% of the available power supply range is wasted. This is a significant limitation especially if the opamp is intended to be a general

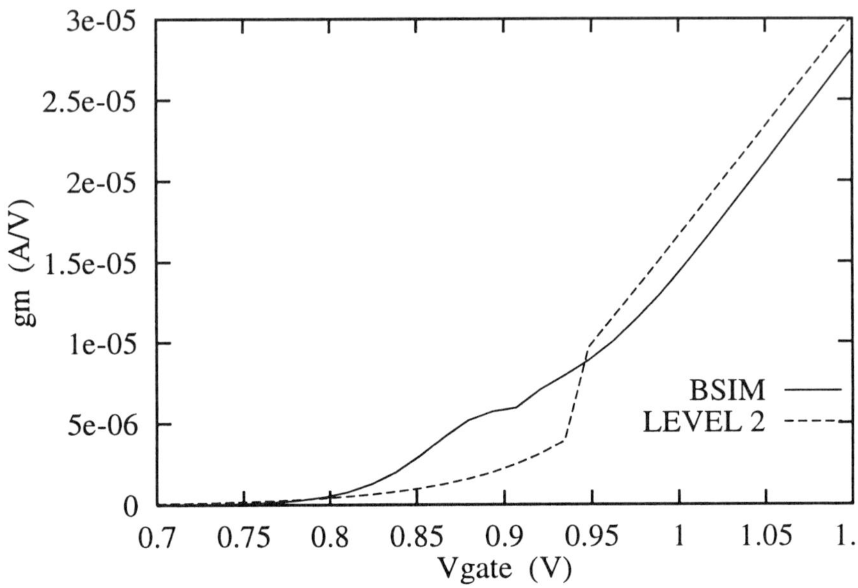

Figure 2.2: Comparison of BSIM and LEVEL 2 models for simulating g_m.

purpose one, *e.g.*, used as a unity gain buffer.

The output swing is also decreased although not as severely as the input stage. However, we can no longer use such circuit technique as a low output resistance source follower in the design of low-voltage opamps. In the following, we will consider the effect of reduced power supply voltage on the large signal behavior of simple but very useful commonly-used analog building blocks.

2.2.1 Input Stage: A CMOS Differential Pair

An n-channel differential pair consisting of transistors M_1 and M_2 with their transconductance parameter equal to K_n is shown in Figure 2.3; it is biased with a current, I_b, provided by M_3. A differential pair is used as an input stage of an opamp. In general the drain terminals of the input transistors are connected to an active load; however, for simplicity they are connected directly to the positive power supply, V_{DD}, in Figure 2.3 for the purpose of determining the input range. As long as the circuit is used with

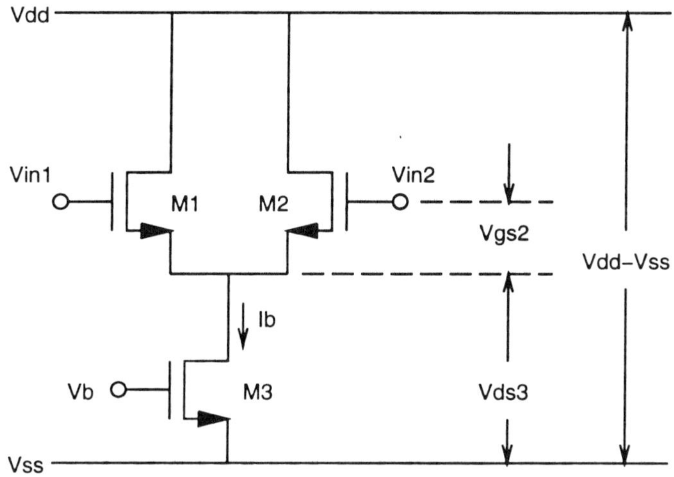

Figure 2.3: An n-channel differential pair.

a negative feedback, we can assume a virtual short between V_{in1} and V_{in2}, and they are equal to the common mode input voltage, V_{CM}. The input common mode range, CMR, is defined as the valid range of V_{CM} which maintains the normal operation of the input stage. That is, each of the input transistors receives about $\frac{1}{2}I_b$ and the transconductance, g_m, equals to $\sqrt{2I_bK_n}$. When CMR is equal to the power supply range, $V_{DD} - V_{SS}$, the input stage has a rail-to-rail input range. For the circuit in Figure 2.3, as long as V_{CM} is sufficiently larger than V_{SS}, all the transistors are in saturation and the gate to source voltage of M_1, and M_2 is given by

$$V_{GS1} = V_{GS2} = \sqrt{\frac{I_b}{2K_n}} + V_{Tn}. \tag{2.5}$$

The minimum drain to source voltage required to keep M_3 in saturation is V_{DS3sat} and this is given by

$$V_{DS3sat} = V_{GS3} - V_{Tn} = \sqrt{I_b/K_3} \tag{2.6}$$

and thus, in order for all the transistors to be operating in the saturation region, V_{CM} must be larger than V_{SS} by at least $V_{GS2} + V_{DS3sat}$. Then the input CMR of this input stage is given by

$$CMR = V_{DD} - V_{SS} - (V_{Tn} + \sqrt{\frac{I_b}{2K_n}} + \sqrt{\frac{I_b}{K_3}}). \tag{2.7}$$

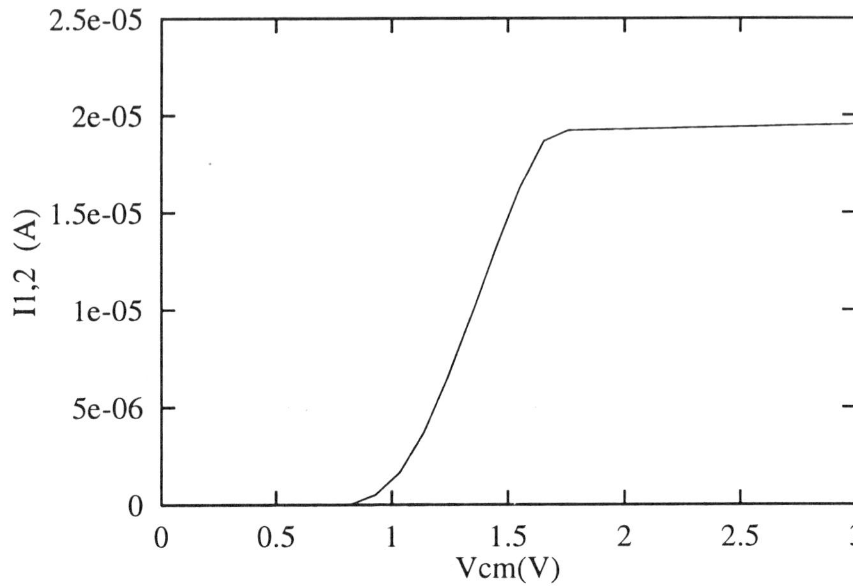

Figure 2.4: Drain current of the simple differential pair as a function V_{CM}.

The term in parentheses can easily be 1.5V or more, so for a 3-V power supply, this differential pair properly operates for only half of the supply range. Figure 2.4 shows the simulation result of the dc-sweep of the differential pair in Figure 2.3. The input common mode voltage is swept between 0 and 3V with $V_{SS} = 0$ and $V_{DD} = 3V$(note that 3.3V with 10% variation has the minimum value of 3V.) The currents in M_1 and M_2 are constant only in the region V_{CM} is larger than 1.5V. As V_{CM} is decreased below 1.5V, transistor M_3 enters the triode region and its drain current, I_b, becomes a function of its drain to source voltage. At the same time V_{GS1} and V_{GS2} are decreasing due to the reduction in I_b. As V_{CM} is decreased further, I_b becomes zero and transistors are completely turned off. Note that while M_3 operates in saturation or triode region in strong inversion, M_1 and M_2 operate in strong or weak inversion in saturation region.

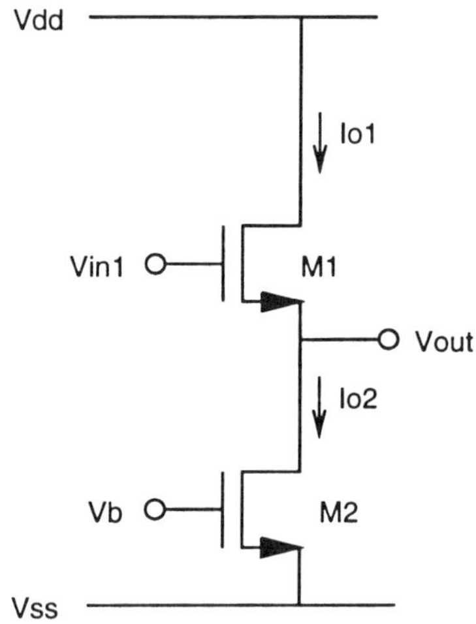

Figure 2.5: A simple CMOS source follower.

2.2.2 Output Stage: A CMOS Source Follower

A source follower shown in Figure 2.5 is often used as an output buffer of larger circuits, including opamps. Ideally, the output of a buffer should follow the input linearly from rail to rail, and provide a small signal gain of one. Further, it should efficiently deliver power to a load with minimum dc power consumption. Biasing for the output transistor M_1 is simply provided by a current sink M_2. Both the input and the output ranges are limited by the threshold voltage of M_1. In order for the circuit to work properly, both transistors must be operating in the saturation region which it is the case as long as

$$V_{in} > V_{Tn} + \sqrt{I_{O1}/K_1} + \sqrt{I_{O2}/K_2} + V_{SS} \tag{2.8}$$

and

$$V_{DD} - (\sqrt{I_{O1}/K_1} + V_{Tn}) > V_{out} > V_{SS} + \sqrt{I_{O2}/K_2} \tag{2.9}$$

are satisfied. Again, for a small power supply voltage, the range for a proper operation is severely limited due to the fact that V_T is not scaled down. The dc transfer curve of the source follower obtained from the simulation is shown in Figure 2.6. The nominal value of the bias current is about $30\mu A$. For V_{in} larger than 1.5V, M_1 receives nominal bias current and thus V_{GS1} is constant. In this region of operation the source follower is operating linear; however, V_{out} can only reach to about 1.5V since V_{GS1} is sitting between

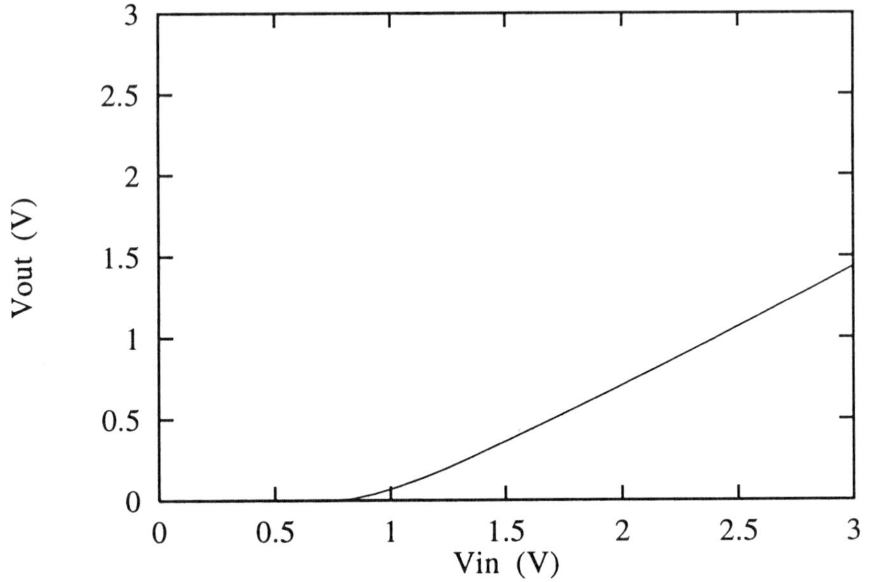

Figure 2.6: dc transfer curve of a CMOS source follower.

V_{DD}(maximum value of V_{CM}) and V_{out}. For V_{CM} less than 1.5V, M_2 is forced into the triode region and the transistors eventually turn off.

We see that a source follower is not at all suited for 3-V operation, and a common source amplifier must be used in order to achieve a large swing at the output. However, the design of an output stage with a common source amplifier is much more complex than with a source follower. In Chapter VI, a class AB controlled source amplifier suitable for low-voltage operation is developed.

2.3 Large Swing Operational Amplifiers

Having input and output terminals that can swing from one power supply rail to the other obviously makes any circuit more attractive, whether the power supply is 10V, 5V, or 3V. Naturally, one can find many publications on opamps with such circuit characteristics both in bipolar[42, 43, 44] and in CMOS[45]-[49] technologies. In these opamps, an n-channel differ-

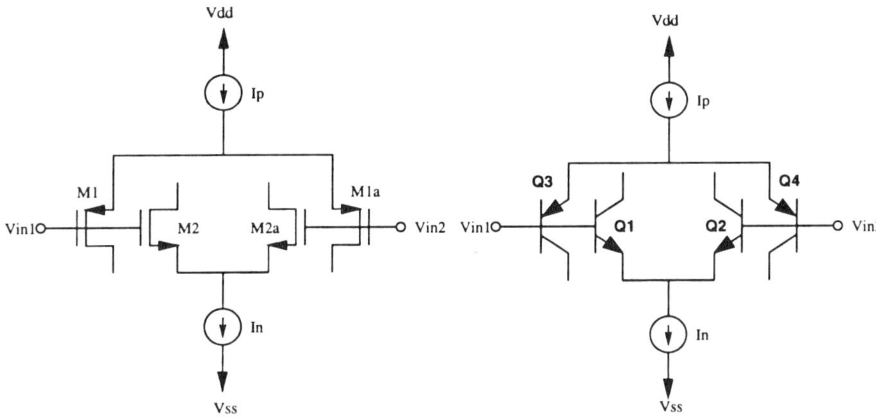

Figure 2.7: Rail-to-rail input stage in CMOS and bipolar implementations

ential pair and a p-channel differential pair are used in parallel as shown in Figure 2.7. This technique enables the input stage to operate rail-to-rail. There are basically three operation regions; when the common mode voltage, V_{CM}, is near the negative power supply, V_{SS}, only the p-channel pair operates. For V_{CM} near the positive power supply, V_{DD}, only the n-channel pair operates. For V_{CM} around mid-rail, both differential pairs operate. That is, at least one of the two differential pairs will be operating for any V_{CM} between the rails.

It is desirable to maintain the nominal performance of the circuit for the entire common mode input range. One of the most important circuit parameters is the transconductance, gm, of the input stage. The total input stage transconductance, g_{mT} of the circuits in Figure 2.7 is given by the sum of the transconductance of the n- and p-channel(or -type) differential pairs, g_{mn} and g_{mp}, respectively. Since there are three regions of operation for the input stage, there are three different regions for the g_{mT}. Figure 2.8 shows the transconductance of the rail-to-rail CMOS input stage as a function of V_{CM}. MOS transistors operating as current sink/source are used to respectively provide I_n and I_p for the simulation since an ideal current source will not turn off regardless of the voltage drop across it. Drain terminals of the n- and p-channel differential pairs are connected to V_{DD} and V_{SS}, respectively, for the simulation since we are only interested in g_m of the differential pairs.

Since the output drain currents of the n- and p-pairs will eventually be

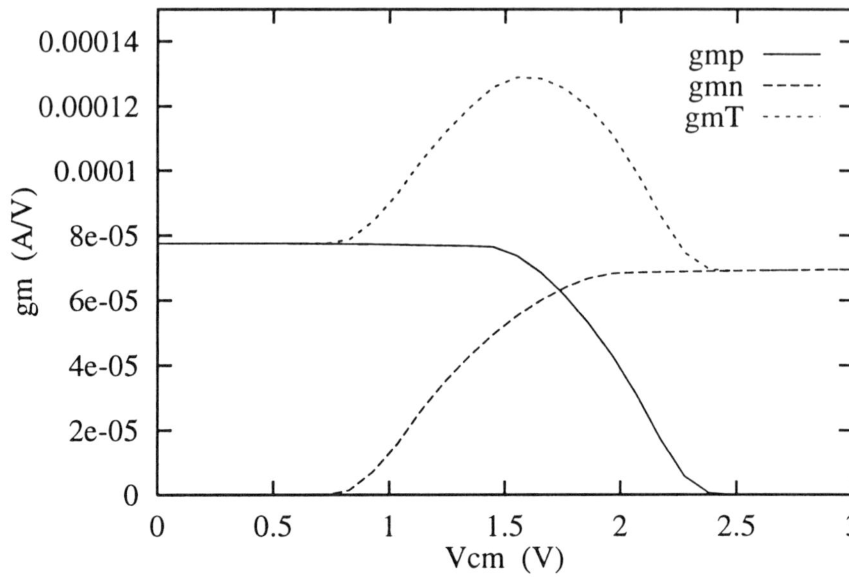

Figure 2.8: Transconductance of a rail-to-rail CMOS input stage as a function of the common mode input voltage.

added up, the total transconductance is given by

$$\begin{aligned} g_{mT} &= g_{mn} + g_{mp} \\ &= \sqrt{2K_n I_n} + \sqrt{2K_p I_p} \end{aligned} \tag{2.10}$$

where K_n and K_p are the transconductance parameters of the n- and p-channel input transistors. If the nominal values of g_{mn} and g_{mp} can be matched, then the ratio of the maximum value of g_{mT}(when both pairs are working) to the minimum value(only one pair is working) is two. The change in g_{mT} with V_{CM} as depicted in Figure 2.8 is not desirable for reasons which are discussed next.

2.3.1 The Unity Gain Frequency, ω_U

Using the opamp model shown in Figure 2.1, we can derive the following opamp frequency characteristics.

$$A_V = g_{mT} g_{mO} R_{O1} R_{O2} \quad (2.11)$$

$$P_1 = -\frac{1}{g_{mO} R_{O1} R_{O2} C_C} \quad (2.12)$$

$$P_2 = -\frac{g_{mO}}{C_{O2}} \quad (2.13)$$

$$Z_1 = \frac{g_{mO}}{C_C} \quad (2.14)$$

where A_V is the dc gain, P_1 and P_2 are the first(or dominant) and the second poles, respectively, and Z_1 is the right hand plane zero. Assuming that P_2 is located at a very high frequency, then the unity gain frequency ,ω_U, or the gainbandwidth product is given by

$$\omega_U = A_V P_1 = \frac{g_{mT}}{C_C}. \quad (2.15)$$

In order to maintain a sufficient amount of phase margin, P_2 must be placed at about two and half times ω_U. In other words, for a given g_{mT} and a second pole, P_2, C_C should be chosen such that $\omega_U = P_2/2.5$ to maximize the opamp gain-bandwidth while maintaining a good opamp performance. All this would be possible provided that g_{mT}, which is a small signal parameter, is constant regardless of the value of the large signal, V_{CM}. However, it was shown in the previous section that for the rail-to-rail CMOS input stage, g_{mT} changes by as much as 100% when V_{CM} is swept from rail to rail. The opamp model of Figure 2.1 was simulated with the following parameter values: $C_{O1} = 0.5pF, C_{O2} = 4pF, C_C = 2pF, R_{O1} = 10M\Omega, R_{O2} = 100K\Omega$,, and $g_{mO} = 250\mu A/V$. Using these values should result in P_2 of about $10MHz$, and this implies that g_{mT} should be such that f_U is about $4MHz$. The simulation results with different values of g_{mT} are shown in Table 2.1 where ϕ_m is the phase margin of the opamp. For $g_{mT} = 50\mu A/V$, $f_U = 3.3MHz$ and this results in a good phase margin. If this was the simple rail-to-rail input stage, then $50\mu A/V$ could either be its maximum or minimum value. In the case that $50\mu A/V$ is the maximum value of g_{mT}, the minimum value would be about $25\mu A/V$ and f_U would go down to $1.8MHz$; and the phase margin will always be more than adequate. In the case that $50\mu A/V$ is the minimum value of g_{mT}, the maximum value might be about $100\mu A/V$, then f_U may be as large as $5.7MHz$. In this case, the phase margin will only be 35°. This example demonstrates that the having a constant-g_m input stage is necessary to optimally compensate the frequency response of the opamp.

The time domain response of the opamp is closely related to the frequency response of the opamp. The settling time, T_S, and the percentage

Table 2.1: Frequency response of the opamp model with various g_{mT} values.

$g_{mT}(\mu A/V)$	$A_V(dB)$	$f_U(MHz)$	$\phi_m(^o)$
25	76	1.8	72
50	82	3.3	57
100	88	5.7	35

overshoot $P.O.$ of a step response are related to f_U and ϕ_m, respectively. These are discussed in a later chapter; however, we must keep in mind that g_{mT} affects many opamp performance criteria and the tradeoffs arising in choosing the value of g_{mT} should always be taken into account.

2.3.2 Harmonic Distortion

The easiest way to see the effect of g_{mT} on the opamp performance is to look at the total harmonic distortion(THD) of the unity gain connected opamp. Consider a signal , $v_{in} = A_{MP}sin(2000\pi t)$, applied to the positive input terminal of an opamp with a direct feedback from the output terminal to the negative input terminal. This input voltage is the common mode input voltage which affects g_{mT}(see Figure 2.8.) Let the gain of the opamp be given by

$$A(v_{in}) = g_{mT}(v_{in})R_O \tag{2.16}$$

where R_O is the output resistance. Then the output voltage is given by

$$v_o = \frac{A(v_{in})v_{in}}{1 + A(v_{in})} = F(v_{in}). \tag{2.17}$$

This can be expanded into

$$v_o = F(0) + F'(0)v_{in} + F''(0)\frac{v_{in}^2}{2} + ... \tag{2.18}$$

where the coefficients $F'(0), F''(0)/2, etc.$ are used to determine THD, and they can be easily calculated as follows.

$$F(0) = 0 \tag{2.19}$$

$$F'(0) = \frac{A(0)}{1 + A(0)} \tag{2.20}$$

$$F''(0) = \frac{2A'(0)}{(1 + A(0))^2} \tag{2.21}$$

$$... \quad ... \tag{2.22}$$

Note that since the ideal expression for the output voltage is $v_o = v_{in}$, $A(0)$ should be as large as possible. Also, in order to minimize the distortion caused by input voltage dependent opamp gain, $A(0)$ must be much larger than $A'(0)$ and the other higher derivatives; in other words, the more constant $A(vin)$(or $g_{mT}(v_{in})$) becomes, the less distortion the input stage would cause.

To demonstrate the above observation, a simple one stage opamp shown in Figure 2.9 is connected in a buffer configuration(V_{in2} and v_o connected while v_{in} is applied to V_{in1}) and is simulated for THD analysis. The rail-to-rail input stage is the same circuit as shown in Figure 2.7 with its g_{mT} as a function of the common mode voltage as shown in Figure 2.8. The gain stage is constructed with ideal elements so that the distortion in the output signal is caused only by the input stage. With a differential signal $V_{in1} - V_{in2} = V_{in}$, the drain currents of the input stage are given by

$$I_1 = \frac{I_p}{2} - g_{mp}\frac{V_{in}}{2} \tag{2.23}$$

$$I_2 = \frac{I_p}{2} + g_{mp}\frac{V_{in}}{2} \tag{2.24}$$

$$I_3 = \frac{I_n}{2} + g_{mn}\frac{V_{in}}{2} \tag{2.25}$$

$$I_4 = \frac{I_n}{2} - g_{mn}\frac{V_{in}}{2} \tag{2.26}$$

then V_o is given by

$$V_o = AV_{in} = -(I_1 - I_2 - I_3 + I_4)R_O \tag{2.27}$$

$$= -(g_{mn} + g_{mp})R_O V_{in} \tag{2.28}$$

$$= -g_{mT} R_O V_{in}. \tag{2.29}$$

With $R_O = 30M\Omega$ and $C_O = 7PF$, the dc gain and f_U are $40dB$ and $1MHz$, respectively, when the common mode voltage is at mid-rail.

Simulation results of THD are shown in Table 2.2; the load resistance used are $30M\Omega$ and $60M\Omega$. A_{MP} is the amplitude of the sinusoid, and the dc level is set at the mid-rail. For a small A_{MP}, the THD increases with increasing input signal. However, when A_{MP} is sufficiently large such that the input stage goes into the operating region where g_{mT} is flat, THD actually starts to decrease. It can be seen that having a larger gain(by increasing R_O) improves the distortion performance though it is not as effective as making the gain more constant. To obtain a more constant-g_m input stage, V_T of the input transistors is reduced by creating a fictitious set of model parameters. Of course this will not be used to actually design a circuit, but it should be justifiable to use just to demonstrate the concept

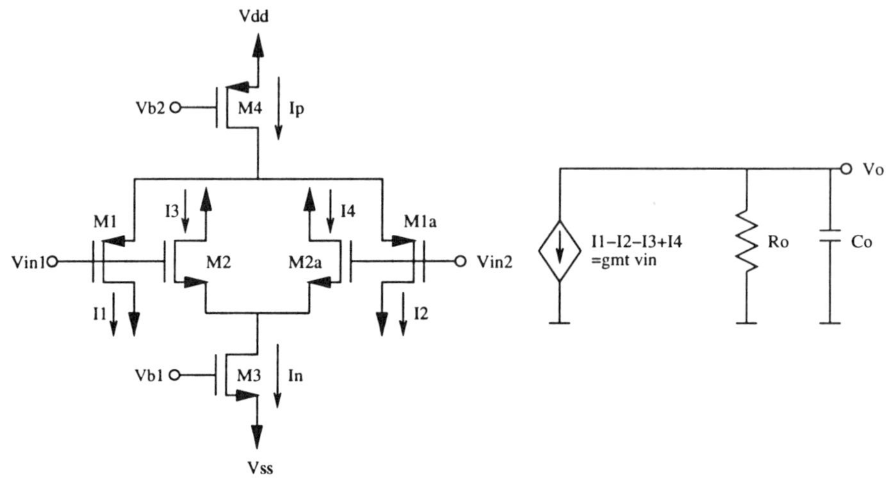

Figure 2.9: A simple single stage opamp

in question. The g_{mT} of the input stage with reduced V_T is shown in Figure 2.10. THD simulation results of the opamp with the reduced V_T are also given in Table 2.2, and we can clearly see the improvement in THD due to a more constant g_{mT}. Unlike the input stage simulated with the normal V_T, THD keeps increasing with increasing A_{MP}. This agrees with the plot of Figure 2.10 in that g_{mT} is flat around the mid-rail and it starts to roll off near the rails. Note that in reality, other stages of an opamp contribute to THD and the benefit of having a constant-g_m input stage is not as large as this example shows; nevertheless the improvement will be significant.

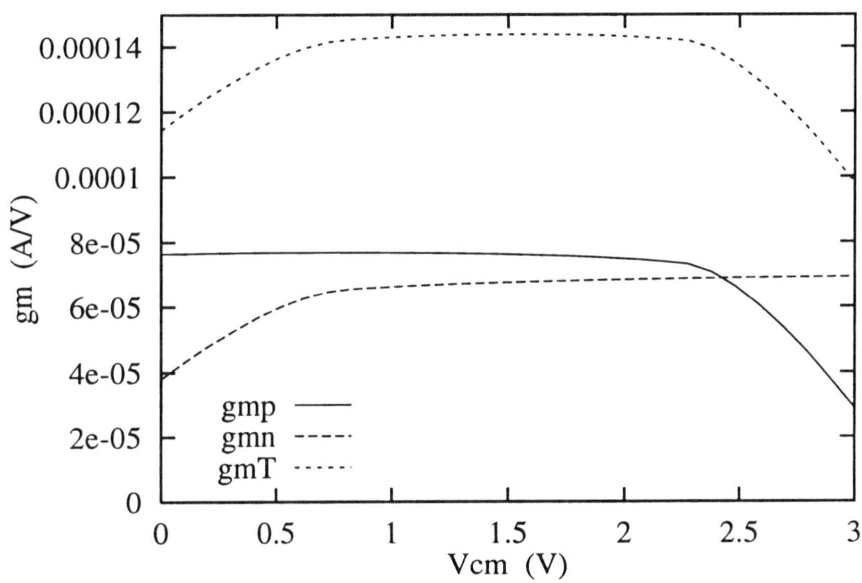

Figure 2.10: Transconductance of a rail-to-rail input stage with reduced V_T.

Table 2.2: THD of the Simple Rail-to-Rail opamp($m = \times 10^{-3}$)

$A_{MP}(V)$	THD(%) $R_O = 30M\Omega$ Normal V_{Ts}	THD(%) $R_O = 60M\Omega$ Normal V_{Ts}	THD(%) $R_O = 30M\Omega$ Reduced V_{Ts}	THD(%) $R_O = 60M\Omega$ Reduced V_{Ts}
0.1	26m	13m	.30m	.15m
0.2	43m	22m	.59m	.30m
0.3	59m	30m	.90m	.45m
0.4	79m	40m	1.3m	.63m
0.5	.10	53m	1.7m	.83m
0.6	.13	67m	2.1m	1.1m
0.7	.17	86m	2.7m	1.4m
0.8	.20	.10	3.6m	1.8m
0.9	.21	.10	7.0m	3.5m
1.0	.19	97m	13m	6.5m
1.1	.18	89m	21m	10m
1.2	.16	81m	29m	15m
1.3	.15	74m	40m	20m
1.4	.13	67m	52m	26m
1.5	.12	62m	66m	33m

Chapter 3

Constant-g_m Input Stages, $K_n = K_p$

When the input stage is constructed using bipolar transistors, it is not very difficult to make g_{mT} constant because the transconductance of a bipolar transistor is proportional to the collector current. Hence the only task needed is to keep the sum of the collector currents of the input pairs constant, *i.e.*, when one of the differential pairs is turned off, the bias current is redirected to the pair which is operating[44].

This is not so simple, however, for the MOS case because the transconductance of a MOS transistor is proportional to the square root of the drain current. Furthermore, g_m also depends on the type of transistor used, *i. e.*, n- or p-channel. A solution was suggested recently[37, 38, 50, 51] under an assumption that an n-channel MOSFET can be matched to a p-channel MOSFET. Validity of this assumption is investigated at the end of this chapter.

In this chapter, circuit techniques that have been used recently[37, 38, 50, 51]to obtain a constant-g_m input stage are discussed. These techniques and the new techniques presented in later chapters, are the only existing solutions leading to constant-g_m input stages. For simplicity, all the n- and the p-channel MOS transistors are assumed to have the transconductance parameter K_n and K_p, respectively, in the analysis of the circuits.

3.1 Constant-g_m Input Stage Using Current Switches

The total transconductance of an input stage with n- and p-channel differential pairs placed in parallel is given by

$$g_{mT} = g_{mn} + g_{mp} \tag{3.1}$$
$$= \sqrt{2K_nI_n} + \sqrt{2K_pI_p} \tag{3.2}$$

and it was shown in the previous chapter that when I_n and I_p are supplied by constant current sources, g_{mT} is not constant as the common mode input voltage, V_{CM}, is changed between the rails. The reason that g_{mT} is not constant is that I_n and I_p are forced to change from their nominal values independent of each other. To keep g_{mT} constant, I_n must increase when I_p decreases, or vice versa. If we can assume that n-channel transistors can be matched to p-channel transistors, *i.e.*, $K_n = K_p = K$, (3.2) can be rewritten as follows.

$$g_{mT} = \sqrt{2K}(\sqrt{I_n} + \sqrt{I_p}) \tag{3.3}$$

Then $\sqrt{I_n} + \sqrt{I_p}$ must be kept constant to obtain a constant-g_m input stage.

If we assume that I_n and I_p are either at their nominal value, I_{ref}, or completely turned off, current switches can be used to keep (3.3) constant. A constant-g_m input stage realized using such technique is shown in Figure 3.1[37]. Current switches M_3 and M_4 are biased with reference voltages V_{b1} and V_{b2}. Under the nominal condition,*i.e.*, when V_{CM} is near the mid-rail so that both differential pairs are operating in the saturation region, switches M_3 and M_4 are turned off and $I_p = I_n = I_{ref}$, then

$$g_{mT} = g_{mn} + g_{mp}$$
$$= 2(\sqrt{I_{ref}})\sqrt{2K} \tag{3.4}$$

When V_{CM} is increased toward V_{DD}, V_b is also increased and the source to gate voltage of M_3 increases. This will turn on M_3 and I_{ref} flows into M_3 instead of M_1 and M_{1a}; at this point, V_b becomes independent of the input voltages because M_1 and M_2 are turned off. M_4 is turned off due to the increase in V_a and hence the total current supplying the n-channel differential pair is $4I_{ref}$. At this point the p-channel pair is turned off and only the n-channel pair contributes to the total transconductance. Then

$$g_{mT} = g_{mn}$$
$$= (\sqrt{4I_{ref}})\sqrt{2K} \tag{3.5}$$

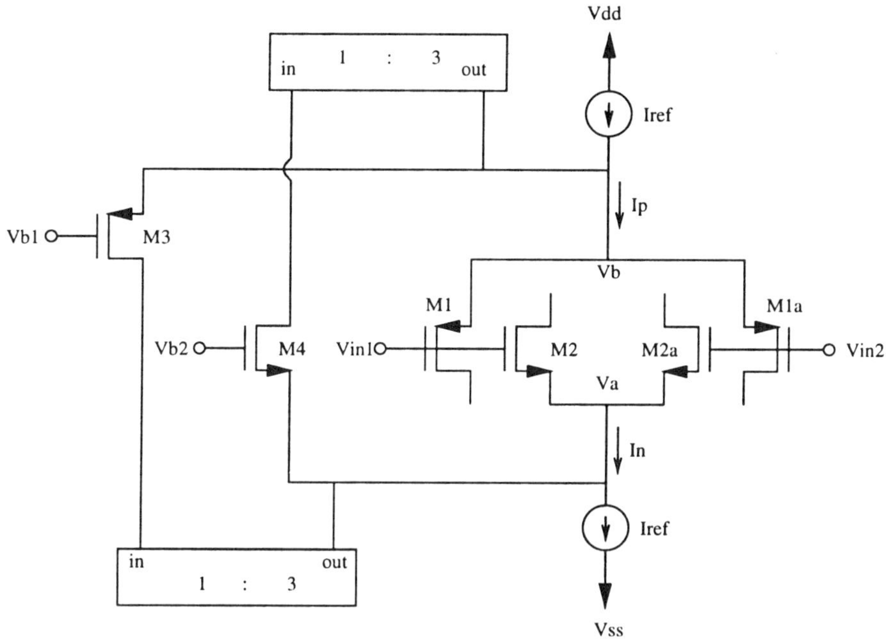

Figure 3.1: A constant-g_m input stage using current switches

which is the same as (3.4). The opposite happens when V_{CM} is reduced toward V_{SS} and the n-channel pair is turned off. V_a and V_b are also reduced and then M_4 is turned on while M_3 is turned off. Then I_{ref} flows through M_4 and with the 3:1 current mirror, the p-channel pair receives $4I_{ref}$. Thus,

$$\begin{aligned} g_{mT} &= g_{mp} \\ &= (\sqrt{4I_{ref}})\sqrt{2K} \end{aligned} \tag{3.6}$$

and g_{mT} is given by $g_{mT} = \sqrt{8I_{ref}K}$ whether one or both differential pairs are operating. When using this circuit, the bias voltages for the switches must be chosen carefully, such that both switches are never turned on at the same time. If both switches are on, a current path consisting of M_3, M_1, M_2, and M_4 is formed; this loop will carry currents which are independent of the rest of the circuit and constant-g_m characteristics will not be achieved.

As we saw in Figure 2.8, transistors do not turn on or off abruptly, rather, the transition region between the fully-on and the fully-off states is quite wide. This implies that there will be regions where a differential pair is receiving currents somewhere between I_{ref} and $4I_{ref}$, not precisely I_{ref}

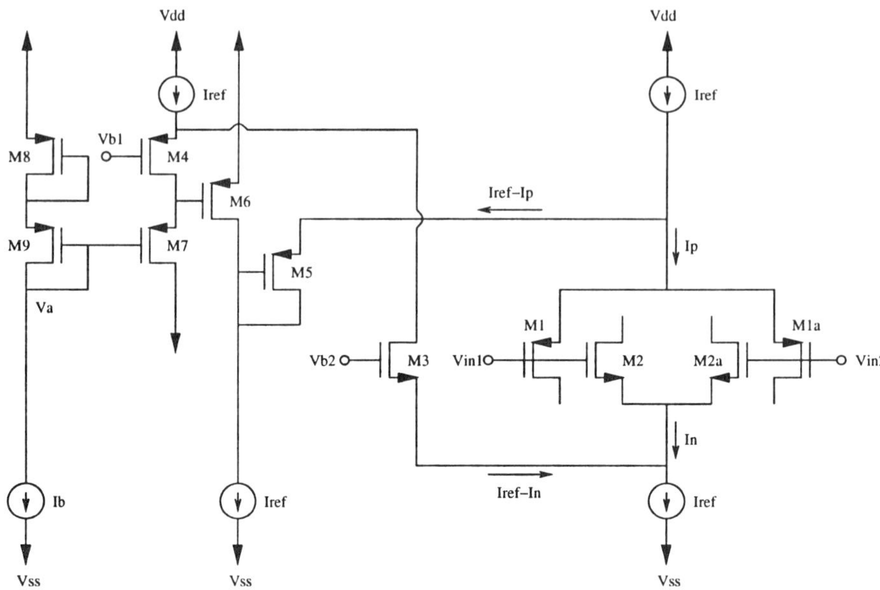

Figure 3.2: A constant-g_m input stage using square-root circuit.

or $4I_{ref}$; then there will be regions where g_{mT} is not so constant. A way to make g_{mT} more constant is to use a square-root circuit as described in the next section.

3.2 Constant-g_m Input Stage Using Square-Root Circuit

In this section, two constant-g_m input stages using the square-root circuit are discussed [37, 38]. We are still trying to implement (3.3); however, I_n and I_p are assumed to vary continuously with V_{CM}. The first circuit is shown in Figure 3.2[37]; the circuit operation is explained as follows. A constant bias current, I_b, biases M_8 and M_9, and the drain voltage of M_9 is constant and given by

$$V_a = V_{DD} - 2|V_{Tp}| - 2\sqrt{\frac{I_b}{K_p}}. \tag{3.7}$$

The sum of the source to gate voltages of M_6 and M_7 equals $V_{DD} - V_a$ and hence it is constant, *i.e.*,

$$V_{SG6} + V_{SG7} = V_{DD} - V_a \tag{3.8}$$

and it can be rewritten in terms of the drain currents as follows.

$$V_{SG6} + V_{SG7} = 2|V_{Tp}| + \sqrt{\frac{I_6}{K_p}} + \sqrt{\frac{I_7}{K_p}} \tag{3.9}$$

where I_6 and I_7 are given by

$$\begin{aligned} I_6 &= I_{ref} - I_5 \\ &= I_{ref} - (I_{ref} - I_p) \\ &= I_p \end{aligned} \tag{3.10}$$

and

$$\begin{aligned} I_7 &= I_{ref} - I_3 \\ &= I_{ref} - (I_{ref} - I_n) \\ &= I_n, \end{aligned} \tag{3.11}$$

respectively. Using these relations we can show that the following is true and a constant-g_m is achieved.

$$\sqrt{I_p} + \sqrt{I_n} = 2\sqrt{I_b}. \tag{3.12}$$

The second implementation of a constant-g_m input stage using a square-root circuit is shown in Figure 3.3. The operation of the circuit is explained as follows. First, notice that the current mirror in the lower left corner maintains the sum of I_3, I_6, and I_8 equal to the sum of I_4, I_5, and I_7, *i.e.*,

$$I_6 + I_8 = I_4 + I_7. \tag{3.13}$$

Note that I_3 and I_5 are equal and thus cancelled. The gate terminal of M_7 is fixed at V_b whereas the gate terminal of M_8, V_C, follows the common mode input voltage. So then, when the common mode voltage is decreasing and the n-channel differential pair is contributing less to g_{mT}, the current in M_8 increases while the current in M_7 decreases. However, I_4 must increase in order to maintain (3.13). Since I_4 equals I_p, the increase in this current increases the contribution of the p-channel differential pair to g_{mT} and compensates for the smaller g_{mn}. The opposite happens when V_C is increased with increasing V_{CM}.

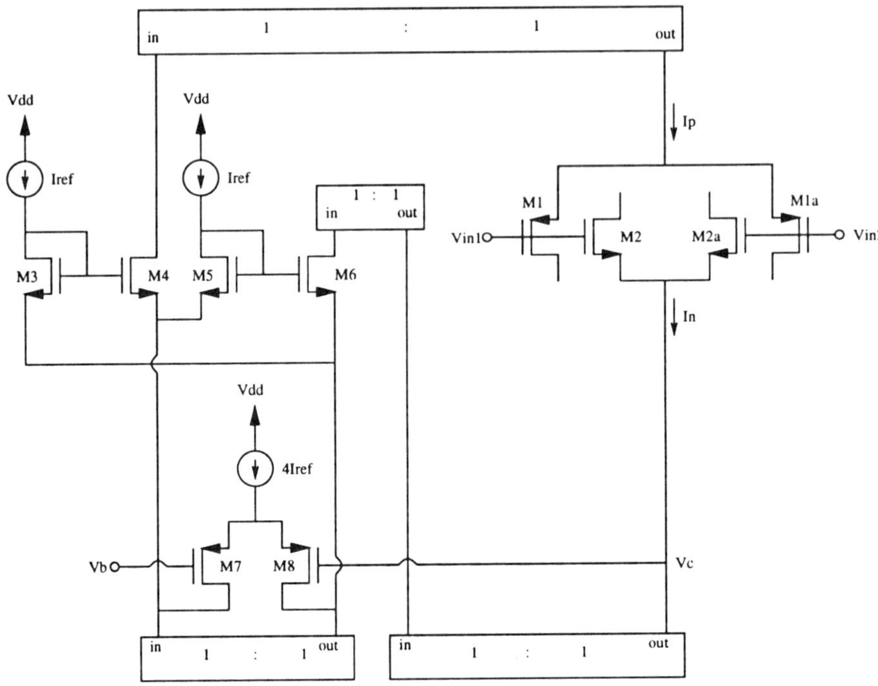

Figure 3.3: An alternate implementation of a constant-g_m input stage using square-root circuit.

The main portion of the circuit consists of transistors M_3, M_4, M_5, and M_6. Summing the gate to source voltages of these transistors, we see that

$$V_{GS3} + V_{GS5} = V_{GS4} + V_{GS6} \tag{3.14}$$

is always true. Then assuming that all transistors are in saturation, (3.14) can be rewritten as

$$\sqrt{I_3} + \sqrt{I_5} = \sqrt{I_4} + \sqrt{I_6}. \tag{3.15}$$

Substituting I_{ref}, I_p, I_{ref}, and I_n for I_3, I_4, I_5, and I_6 into (3.15), we get

$$\sqrt{I_n} + \sqrt{I_p} = 2\sqrt{I_{ref}}. \tag{3.16}$$

Thus, using I_p and I_n as the bias currents for the differential pairs, g_{mT} of the input stage becomes constant.

3.3 Practical Considerations

As mentioned, constant-g_m input stages using CMOS technology is a fairly new topic and existing methods which have been published were discussed in the previous sections. The concepts are simple and the methods are solid and it seems that they can be easily applied in a practical implementation of the circuit utilizing these methods. However, they rely on an assumption which might seem difficult to achieve with good accuracy.

The assumption was that n- and p-channel transistor can be matched, that is $K_n = K_p$ should be achieved by properly sizing the MOS transistors. Since

$$K_n = \frac{1}{2}\mu_n C_{OX}\left(\frac{W}{L}\right)_n \tag{3.17}$$

and

$$K_p = \frac{1}{2}\mu_p C_{OX}\left(\frac{W}{L}\right)_p, \tag{3.18}$$

with prior knowledge of the ratio of the electron mobility to the hole mobility, μ_n/μ_p, $K_n = K_p$ should be made possible by correctly choosing $(W/L)_n$ and $(W/L)_p$. The ratio μ_n/μ_p is usually assumed to be between 2 and 3, and typically a ratio of 2.5 is used for designs which does not require precise matching between opposite transistor types. A good example, of course, is the design of digital CMOS integrated circuits which rely on good noise margin making precise matching unnecessary. However, for the design of a constant-g_m input stage, if the design was carried out assuming $\mu_n/\mu_p = 2.5$ while the actual ratio was 2 or 3, this deviation itself will result in the g_m variation of about 10%($\sqrt{2/2.5} = 0.9$ or $\sqrt{3/2.5} = 1.1$.) This error plus the error which is inherent to the circuit caused by second order effects of transistors can be significantly large to make the circuits presented in this chapter not so attractive.

In general, different processes will have different values of μ_n/μ_p ratio, furthermore, even in the same process the ratio can change considerably from a run to run. To acquire more realistic information on the variation of the μ_n/μ_p ratio, several process parameter sets from venders used by the MOSIS service were obtained. The μ_n/μ_p ratios obtained from the following processes: 1.2μm HP, 1.2μm HP-NID, 2.0μm Orbit, and 2.0μm VTI, are plotted in Figure 3.4. The variations in the ratio can be as much as 30% for a particular process, that is if the design was carried out using the median value of the ratio, the final value may have a $\pm$15% deviation. The circuit in Figure 3.3 was simulated and the result is shown in Figure 3.5 with the curve labeled 'Original μn'. Transistor sizes of the differential pairs were chosen such that g_{mT} with $V_{CM} = 0$ and $V_{CM} = 3V$ is the

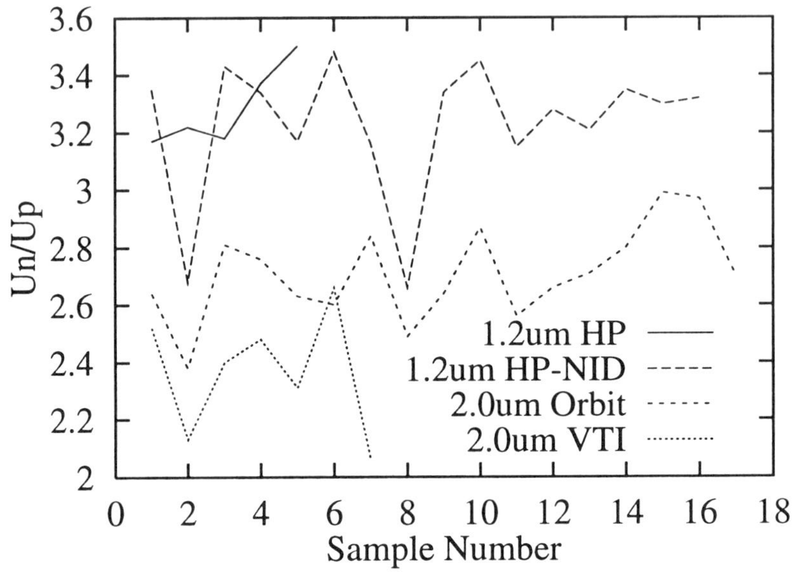

Figure 3.4: Ratios of μ_n to μ_p for different runs and processes.

same. This results in g_{mT} deviation of about 10% as V_{CM} is swept from 0 to 3V. The same circuit was simulated with the n-channel transistor parameter set modified so that μ_n is 15% less and 15% more than the original value; curves are labeled as '15% less' and '15% more'. The modification of the parameter sets results in an invalid transistor model and it is done here only for the purpose of demonstration. Increasing and decreasing μ_n results in g_{mT} deviation of 20% and 8%, respectively. Deviation in g_{mT} caused by the difference in the predicted and the actual μ_n/μ_p will vary depending on a particular run of a particular process. Nevertheless, making the g_{mT} deviation independent of μ_n/μ_p will significantly improve the robustness of the constant-g_m input stage and also ease the circuit design procedure by eliminating the task of guessing the value of μ_n/μ_p for the next process to be used. The use of statistical models which account for transistor mismatches may be incorporated into the design procedure; however, existing statistical models[52, 53] account only for mismatches between devices of the same type. Since the processing steps for n- and p-channel transistors are different, development of statistical models may be extremely difficult.

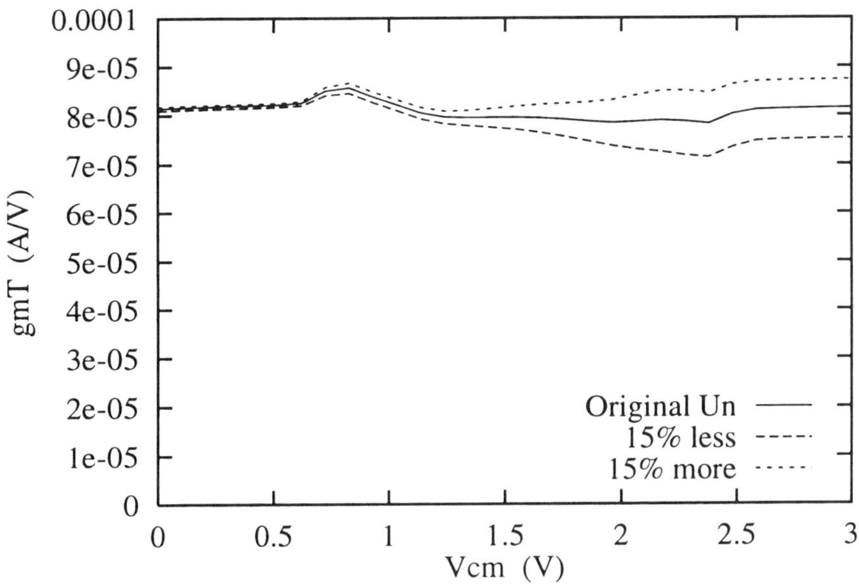

Figure 3.5: Simulation results of the second constant-g_m input stage using square-root circuit with different μ_n values.

Furthermore, while the use of the statistical models may enable us to better predict the mismatches, it helps very little to minimize the effect of the mismatches. An innovative approach would be to investigate circuit design solutions which circumvent the $K_n - K_p$ matching requirement. This establishes the motivation behind the work. In the following chapters new circuit design solutions are presented.

Chapter 4

Robust Bias Circuit Techniques

In this chapter, design solutions which lead to the development of a new rail-to-rail CMOS input stage with a constant-g_m and without the need for matching n-channel transistors to p-channel transistors are presented[54].[1] The design goals are the same as discussed in Chapter III: to develop a bias circuit which produces bias currents I_n and I_p in such a way that the total transconductance of the n- and p-channel differential pairs is constant over the entire common mode range . That is,

$$\begin{aligned} g_{mT} &= g_{mn} + g_{mp} \\ &= \sqrt{2K_n I_n} + \sqrt{2K_p I_p} \end{aligned} \tag{4.1}$$

must be maintained constant without having to factor out $\sqrt{K_n}$ and $\sqrt{K_p}$, *i.e.* $\sqrt{K_n} \neq \sqrt{K_p}$.

The block diagram we propose for a constant-g_m input stage is shown in Figure 4.1. Bias current I_{nmax} is supplied to the constant-g_m bias circuit which contains a MOS circuit that keeps $g_{mn} + g_{mp}$ constant. It also contains a circuit that produces I_{pmax}; the purpose for I_{pmax} and its relation to I_{nmax} are discussed later in this chapter. The constant-g_m bias circuit produces I_p that is fed into the p-channel differential pair. The actual value of I_p being used by the p-channel differential pair is monitored and is fed back into the bias circuit as the common mode input voltage, V_{CM}, is

[1] S. Sakurai and M. Ismail, "Constant transconductance bias circuit and method", U.S. Patent pending No. 08/111,708, filed August 25, 1993. First draft March 1993 submitted to the Semiconductor Research Corporation, North Carolina.

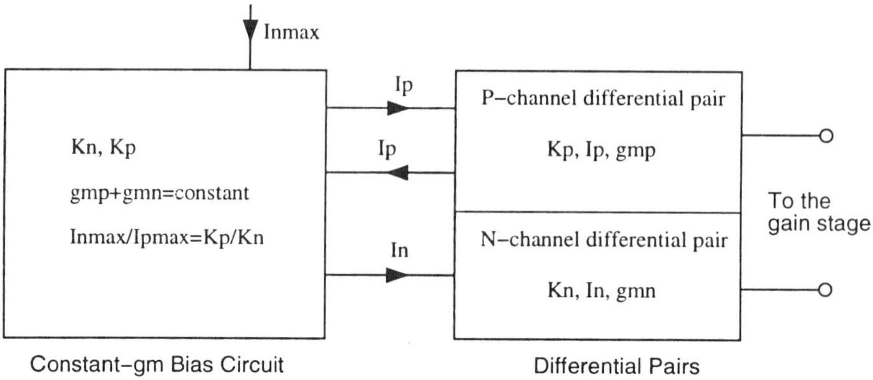

Figure 4.1: The block diagram of a constant-g_m input stage.

varied. This information is used to generate I_n, that biases the n-channel differential pair, so that g_{mT} is constant.

Note that the roles of I_n and I_p can easily be interchanged in the operation, *i.e.*, I_n can be monitored and used to generate I_p. In the following, all n- and p-channel transistors are assumed to have transconductance parameter K_n and K_p, respectively, unless otherwise specified.

4.1 New Circuits for Constant-g_m Input Stages

In this section, CMOS subcircuits which directly realize the relation given in (4.1) are introduced.

Constant-g_m bias circuit 1 is shown in Figure 4.2. The input to the circuit is I_p; as mentioned earlier this is the monitored value of the current which is biasing the p-channel differential pair. The output of this circuit is I_n and this biases the n-channel differential pair. The main portion of the circuit consists of M_3 and M_4. Transistor M_3 is a diode connected n-channel transistor and its source terminal is grounded. Transistor M_4 is a p-channel transistor and its source is biased with a constant voltage V_C. The drain current of M_4 is fed into a current mirror and I_n is made available to the n-channel differential pair through M_6. Since V_C is constant, the sum of the gate to source voltages of M_3 and M_4 is always constant. Assuming that all transistors are operating in the saturation region in the strong inversion region, we have

$$V_C = V_{SG4} + V_{GS3}$$

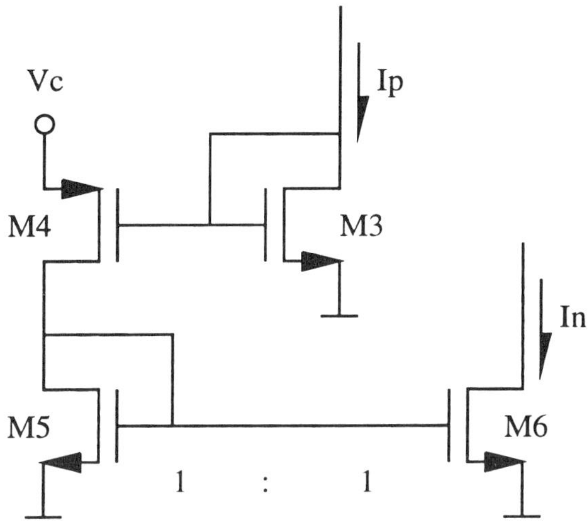

Figure 4.2: A new constant-g_m bias circuit using a bias voltage, V_C, for its reference.

$$= \quad V_{T4} + \sqrt{\frac{I_4}{K_p}} + V_{T3} + \sqrt{\frac{I_3}{K_n}}. \tag{4.2}$$

However, since $I_3 = I_p$ and $I_4 = I_n$, (4.2) can be rewritten as

$$V_C - V_{T3} - V_{T4} = \sqrt{I_p/K_n} + \sqrt{I_n/K_p}. \tag{4.3}$$

Multiplying the above by $\sqrt{K_n K_p}$, we have the following.

$$\sqrt{K_n K_p}(V_C - V_{T3} - V_{T4}) = \sqrt{K_n I_n} + \sqrt{K_p I_p} \tag{4.4}$$

Since the source terminals of M_3 and M_4 are connected to fixed voltages, the threshold voltage of these transistors are constant to the first order, and the left side of (4.4) is constant. Since the currents I_p and I_n in the constant-g_m bias circuit are the same currents that are used to bias the input differential pairs, the right side of (4.4) is the total transconductance, g_{mT}, of the input stage(off by a factor of $\sqrt{2}$). Therefore, the circuit shown in Figure 4.2 can be used to construct a constant-g_m input stage without matching n-channel MOS transistors to p-channel MOS transistors.

The actual values of V_{T3} and V_{T4} are difficult to predict. Since V_C must be smaller than 3V and the sum of V_{T3} and V_{T4} are approximately 2V, a small change in V_C, V_{T3}, and V_{T4} can cause a significant change in g_{mT}.

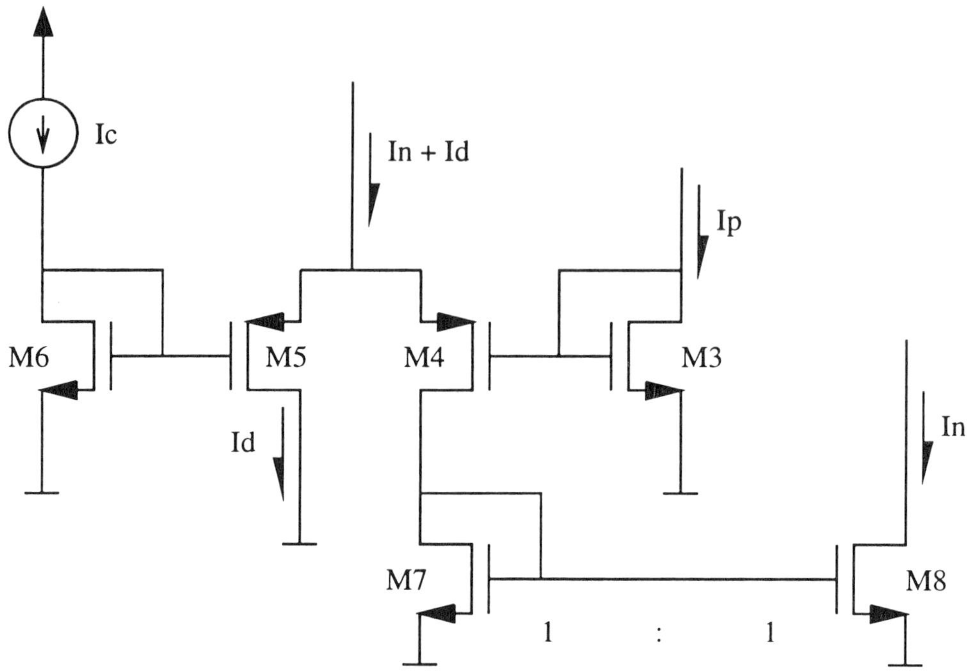

Figure 4.3: A new constant-g_m bias circuit using bias currents , I_c and I_d, for its references.

Thus we should at least get rid of the dependence of g_{mT} on the threshold voltages of either transistor types.

Constant-g_m circuit 2 is shown in Figure 4.3. Again, the current I_p being fed to the transistor M_3 is the input to the bias circuit and I_n which is sourced by M_8 is the output. Transistors M_3 through M_6 form a circuit which maintains the relation between I_p and I_n for a constant g_m. The drain current in M_4 is I_n and this current is fed back into the source terminals of M_4 and M_5; a constant current I_d is also fed into this terminal and thus the current in M_5 equals to I_d. A constant current I_c is fed into a diode connected M_6, then since both M_5 and M_6 are carrying constant currents, their gate to source voltages are constant. This establishes a constant voltage node at the source terminal of M_4 as it was in the constant-g_m bias circuit 1 shown in Figure 4.2. Since the source terminals of M_3 and M_6 are both grounded, the following is always true.

$$V_{SG5} + V_{GS6} = V_{SG4} + V_{GS3} \tag{4.5}$$

Assuming that all the transistors are operating in the saturation region(in strong inversion), (4.5) can be written as

$$\sqrt{\frac{I_c}{K_n}} + V_{T6} + \sqrt{\frac{I_d}{K_p}} + V_{T5} = \sqrt{\frac{I_n}{K_p}} + V_{T4} + \sqrt{\frac{I_p}{K_n}} + V_{T3}. \quad (4.6)$$

Since the p-channel transistors M_4 and M_5 have the same source terminals, the body effect on these transistor should cancel to the first order, that is, we can assume $V_{T4} = V_{T5}$. This is true for the n-channel transistors M_3 and M_6 and then $V_{T3} = V_{T6}$ is also assumed. Using these facts, (4.6) becomes

$$\sqrt{\frac{I_c}{K_n}} + \sqrt{\frac{I_d}{K_p}} = \sqrt{\frac{I_n}{K_p}} + \sqrt{\frac{I_p}{K_n}}. \quad (4.7)$$

Multiplying both sides of (4.7) by $\sqrt{2K_nK_p}$, we get

$$\sqrt{2I_cK_p} + \sqrt{2I_dK_n} = \sqrt{2I_nK_n} + \sqrt{2I_pK_p} \quad (4.8)$$

where the right side of the equation equals g_{mT} of the input differential pairs. The minimum voltage drop required to operate this circuit is the sum of the gate to source voltages of M_5 and M_6 plus the minimum saturation voltage for the current source which provides $I_n + I_d$ into the sources of M_4 and M_5.

Constant-g_m circuit 3 is shown in Figure 4.4. Note that the operation is similar to the constant-g_m bias circuit 2 which is shown in Figure 4.3, and only the main part that achieves the constant g_m is shown. Starting from the gate terminal of M_3, we can sum the gate to source voltages of each transistor forming a loop as follows.

$$V_{GS3} - V_{GS6} + V_{SG4} - V_{SG5} = 0 \quad (4.9)$$

or

$$V_{SG5} + V_{GS6} = V_{SG4} + V_{GS3} \quad (4.10)$$

as in (4.5). Again if we assume that all transistors are operating normally, we can show that this circuit achieves the relation given in (4.6). Unlike the circuit shown in Figure 4.3 the source terminals of the same type transistors are not connected to each other. Then V_{T3} and V_{T4} are only approximately equal to V_{T6} and V_{T5}, respectively. Then instead of (4.8), the following should be used to describe the operation of the circuit shown in Figure 4.4.

$$\sqrt{2I_cK_p} + \sqrt{2I_dK_n} \approx \sqrt{2I_nK_n} + \sqrt{2I_pK_p}. \quad (4.11)$$

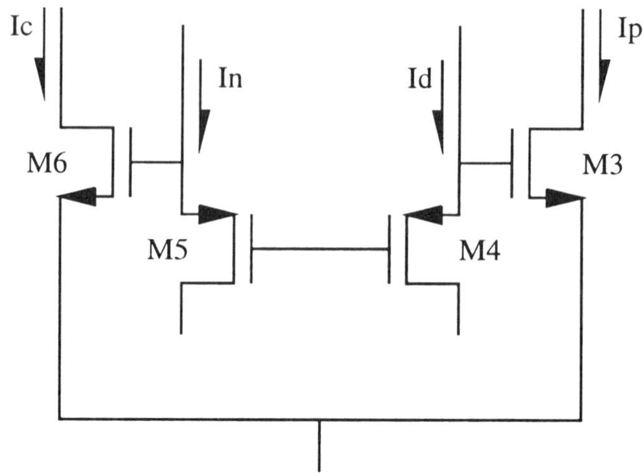

Figure 4.4: An alternate realization of new constant-g_m bias circuit using bias currents.

However, this alternate circuit topology requires only one gate to source voltage drop, either V_{SG5} or V_{GS6} whichever is larger, plus the minimum saturation voltage for the current source. So there is a tradeoff between the accuracy of the constant g_m and the minimum operating voltage required.

In the remainder of this work, constant-g_m bias circuit 2 will be used as the bias circuit for the input differential pairs. This is due to its independence of the transistor threshold voltage and because it is free from body effects. The target minimum power supply voltage is 3V and the minimum voltage required for the operation of the constant-g_m bias circuit 2 is well below this value.

4.2 Current Monitoring Schemes

In the previous section, CMOS circuits that provide the relation $\sqrt{K_n I_n} + \sqrt{K_p I_p} = constant$ were introduced. In all of them, it was assumed that the input to the bias circuit is a copy of the common mode current in the p-channel differential pair and the output of the bias circuit is the bias current for the n-channel pair; however, connections of these circuits to the differential pairs were not shown explicitly. Again, the roles of I_n and I_p can be switched easily but we will keep I_p as the current that needs to be monitored. Figure 4.5 shows the most general presentation of the constant-g_m input stage. Whether the transistor M_P is biased independently or in

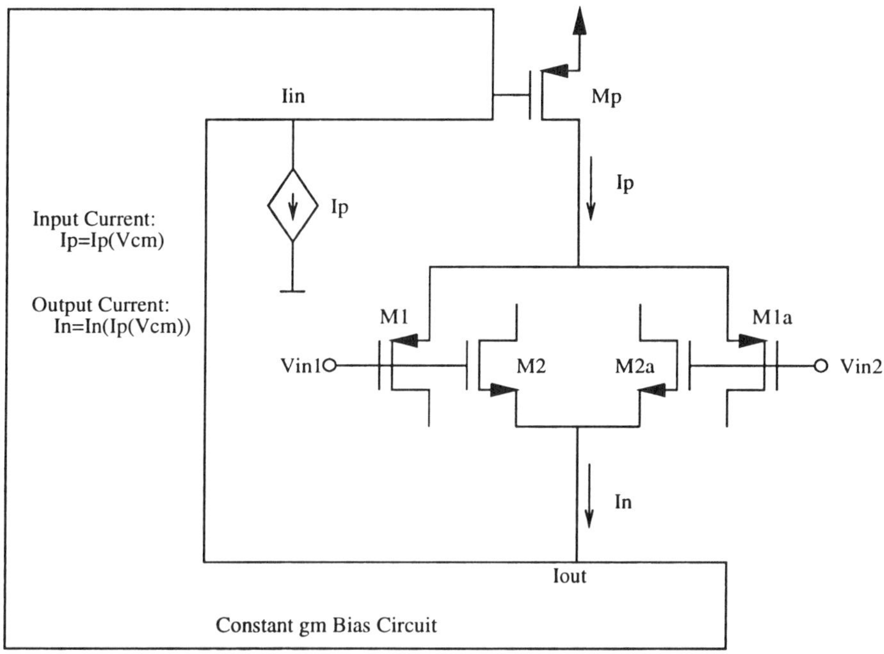

Figure 4.5: A general representation of the constant-g_m input stage consisting of the differential pairs, constant-g_m bias circuit, and current monitor for I_p.

a manner dependent on the constant-g_m bias circuit, its drain current is dependent on the common mode input voltage, V_{CM}. In the following, two schemes for monitoring I_p and feeding it into the constant-g_m bias circuit are discussed.

4.2.1 Monitor 1: Fixed Bias Voltage for M_P

It was shown in Figure 2.8 that using transistors as constant bias current sources results in g_{mn} and g_{mp} which are dependent on V_{CM} but independent of each other. The plot is again shown in Figure 4.6(a) where V_n is defined as the common mode input voltage which results in g_{mn} to be half of its maximum value; the maximum should be obtained when $V_{CM} = 3V$. Similarly, V_p is defined as V_{CM} which results in g_{mp} to be half of its maximum value. In order to have constant-g_m characteristics, we must at least have the following.

$$g_{mp}(V_{CM}=0)=g_{mn}(V_{CM}=3)=g_{mp}(V_{CM}=V_p)+g_{mn}(V_{CM}=V_n), \tag{4.12}$$

i.e., V_p must equal to V_n. Figure 4.6(b) shows transconductance curves which possess such characteristics; the curve for g_{mp} and the location of V_p are an exactly the same as those in Figure 4.6(a). This means that in Figure 4.5, the gate voltage of M_P is independent of V_{CM} and it is constant and the constant-g_m bias circuit is used to move and reshape the g_{mn} curve so that V_n aligns with V_p and the maximum value of g_{mn} is obtained before V_{CM} reaches 3V, or simply, so that the sum of g_{mp} and g_{mn} is constant.

A CMOS circuit which can be used to achieve the g_{mn}-g_{mp} characteristics of Figure 4.6(b) is shown in Figure 4.7. The diode connected transistor M_r, which carries a constant bias current I_{ref}, is the input transistor of a current mirror consisting of M_p, M_q, and M_r. Just as M_p sources current to $M_1 - M_{1a}$ p-channel pair, M_q sources a current to M_3 and M_{3a} whose gates are connected to V_{in1} and V_{in2}. The set of transistors M_3, M_{3a}, and M_q are an exact replica of the transistors M_1, M_{1a}, and M_p so that these two sets behave exactly the same as a function of V_{CM}. Drain terminals of M_3 and M_{3a} are connected to each other and to the I_{in} terminal of the constant-g_m bias circuit. Thus, monitoring the current in the p-channel differential pair and feeding it into the constant-g_m bias circuit is accomplished. The differential input signals applied to M_3 and M_{3a} cancels at their drains and they will not affect the operation of the constant-g_m bias circuit.

The operation of the monitor circuit 1 as a function of V_{CM} is briefly described as follows. When V_{CM} is small enough, transistors M_p and M_q are operating in the saturation region and thus the current mirror is working properly; then $I_p = I_{ref}$. As V_{CM} is increased toward the upper rail, M_p and M_q start to go into the triode region; since their gates are at a fixed voltage, I_p becomes less than I_{ref} and becomes a function of V_{CM}, *i.e.*, we have

$$I_p = \begin{cases} I_{ref} & \text{if } V_{SDp} = V_{SDq} > V_{SDp,sat} = V_{SDq,sat} \\ I_p(V_{CM}) & \text{otherwise} \end{cases}$$

4.2.2 Monitor 2: Actively Biased Voltage for M_P

In the monitor circuit 1, the transistor M_p, which provides a bias current to the p-channel differential pair, is biased with a fixed voltage. An alternate method for monitoring I_p is to actively bias M_p so that both I_n and I_p are always dependent on V_{CM}.

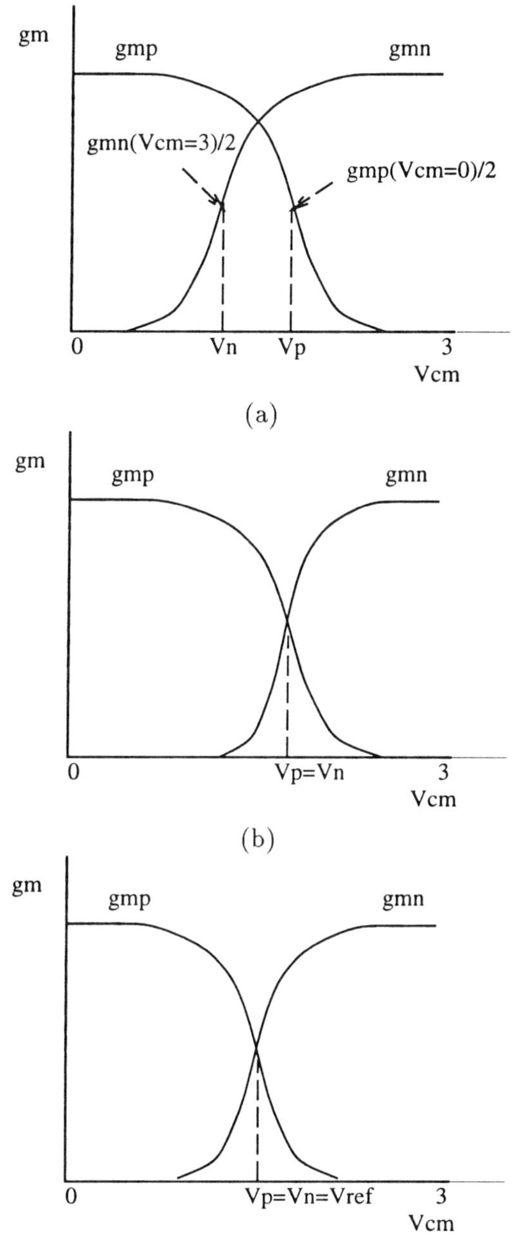

Figure 4.6: Transconductance of the differential pairs: (a) without the constant-g_m bias circuit, (b) with the constant-g_m bias circuit and using monitor 1, (c) with the constant-g_m bias circuit and using monitor 2.

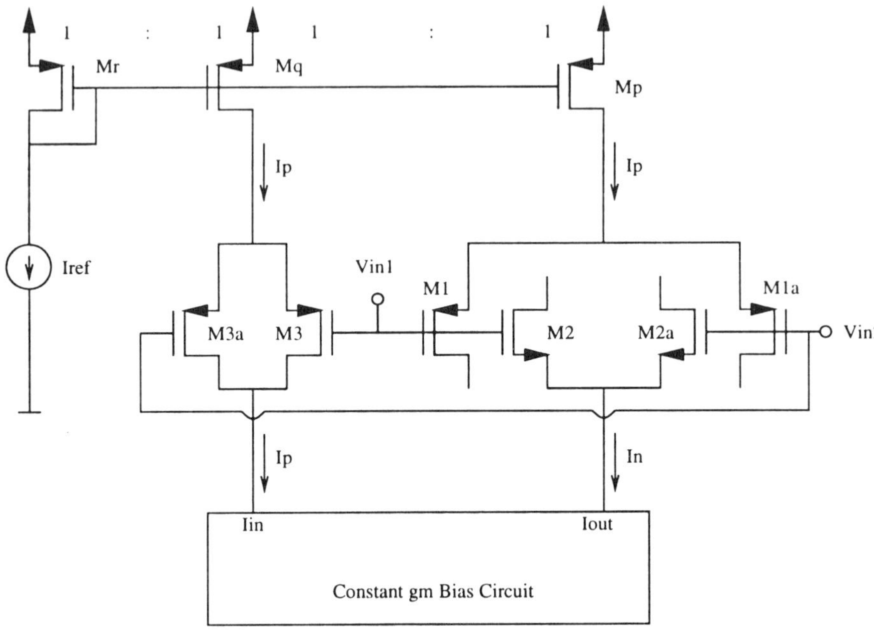

Figure 4.7: A CMOS implementation of monitor 1 which has a current sourcing transistor M_p with fixed bias voltage.

Plots of g_{mn} and g_{mp} which have such characteristics are shown in Figure 4.6(c) in which the point where V_p and V_n meet is externally determined by V_{ref}. Figure 4.8 shows a CMOS circuit that can be used to obtain such characteristics. An n-channel differential pair consisting of M_3 and M_{3a} is placed in parallel with the input differential pairs. The gate terminal of the transistor M_b is biased with a constant voltage V_b, and the source terminal is connected to the source terminals of M_3 and M_{3a}; this node is also connected to a constant current I_{ref}. The drain current of M_b which is labeled as I_p is supplied to the p-channel input differential pair and to the constant-g_m bias circuit through a current mirror consisting of M_r, M_q, and M_p.

As V_{in1} and V_{in2} and hence V_{CM} are increased, a larger portion of I_{ref} is consumed by M_3 and M_{3a}. Then the drain current in M_b reduces and I_p reduces with increasing V_{CM}. This is exactly what we need since the contribution of g_{mp} to the total transconductance should be decreasing with increasing V_{CM}. The value of V_{ref} depends on V_b, I_{ref}, and the sizes of

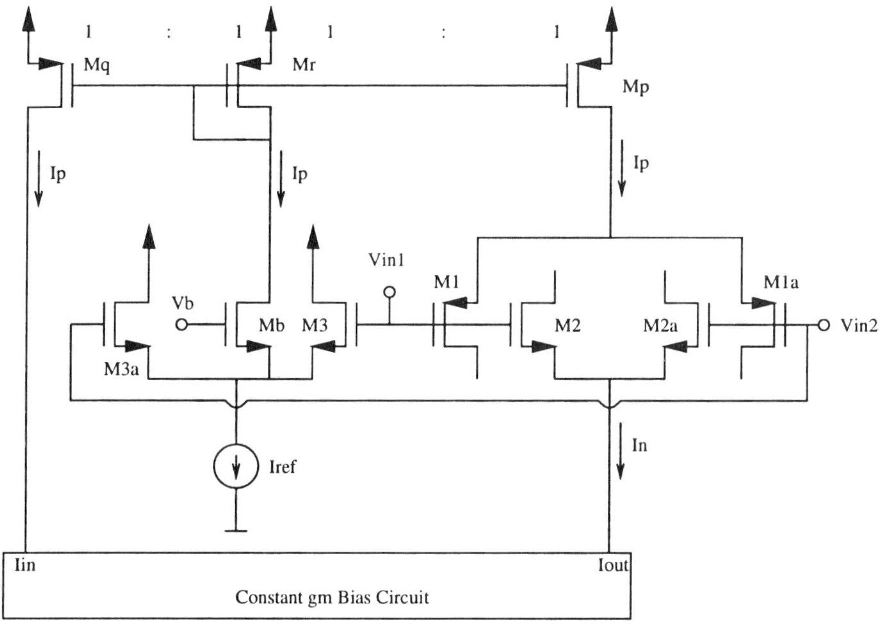

Figure 4.8: A CMOS implementation of monitor 2 which has a current sourcing transistor M_p that is actively biased.

M_3, M_{3a}, and M_b. Assuming that $K_3 = K_{3a}$, we have $I_3 = I_{3a}$, and

$$I_{ref} = 2I_3 + I_b. \tag{4.13}$$

Also, assuming $V_{in1} = V_{in2} = V_{CM}$, we have

$$V_b + V_{GS3} = V_{CM} + V_{GSb} \tag{4.14}$$

or

$$\begin{aligned} V_{CM} &= V_b + \sqrt{I_3/K_3} - \sqrt{I_b/K_b} \\ &= V_b + \sqrt{\frac{I_{ref} - I_p}{2K_3}} - \sqrt{\frac{I_p}{K_b}}. \end{aligned} \tag{4.15}$$

When $V_n = V_p$, g_{mn} and g_{mp} are the same; then if K_n and K_p are approximately the same, V_{CM} should equal V_{ref} when $I_p \approx I_n \approx I_{ref}/4$. Using these observations, (4.15) can be used to determine V_{ref} as follows.

$$\begin{aligned} V_{ref} &\approx V_b + \sqrt{\frac{3I_{ref}}{8K_3}} - \sqrt{\frac{2I_{ref}}{K_b}} \\ &= V_b + \sqrt{2I_{ref}}\left(\sqrt{3/K_3} - \sqrt{2/K_b}\right) \end{aligned} \tag{4.16}$$

Note that if $K_b/K_3 = 2/3$ then $V_{ref} \approx V_b$; also in order to make sure that the current source I_{ref} is operating properly, V_b must be large enough so that

$$V_b > \sqrt{I_{ref}/K_b} + V_{Tb} + V_{DSref,sat} \tag{4.17}$$

where V_{Tb} is the threshold voltage of M_b and $V_{DSref,sat}$ is the minimum voltage required to operate the current source.

The main difference between the two methods used for monitoring I_p is that in monitor circuit 1, the transistor M_p was pushed into the triode region and that resulted in reduction of I_p, whereas in monitor circuit 2, I_p is steered away from the p-channel differential pair before M_p is affected by V_{CM}. Monitors 1 and 2 are simulated without the n-channel differential pairs. Figure 4.9(a) shows the drain current of M_p as a function of V_{CM}; the bias current I_{ref} is set to $20\mu A$. The curve for the monitor 1 is similar to g_{mp} in Figure 2.8(off by a factor of $\sqrt{2K_p}$). Monitor 2 is simulated with $V_b = 1.5V$ and $K_b/K_3 = 2/3$ so that $V_{ref} = 1.5V$. At V_{ref}, I_p should be one-forth of I_{ref}, and we see that I_p for monitor 2 is about $5\mu A$ at $V_{CM} = V_{ref}$ as expected. The source to drain voltage, V_{SDp}, of M_p is shown in Figure 4.9(b) along with the saturation voltage, $V_{SDp,sat}$. For monitor 1, since the gate terminal of M_p is fixed, $V_{SDp,sat}$ is constant and it is about 0.35V; the V_{SDp} curve intersects $V_{SDp,sat}$ at about $V_{CM} = 1.5V$ and we can see that I_p starts to decrease around this point. For monitor 2, since the gate terminal of M_p moves toward V_{DD} with increasing V_{CM}, $V_{SDp,sat}$ also decreases; this starts at about $V_{CM} = 1.1V$ which is the point when I_p starts to decrease. Note that as expected, the V_{SDp} curve never intersects the $V_{SDp,sat}$ curve.

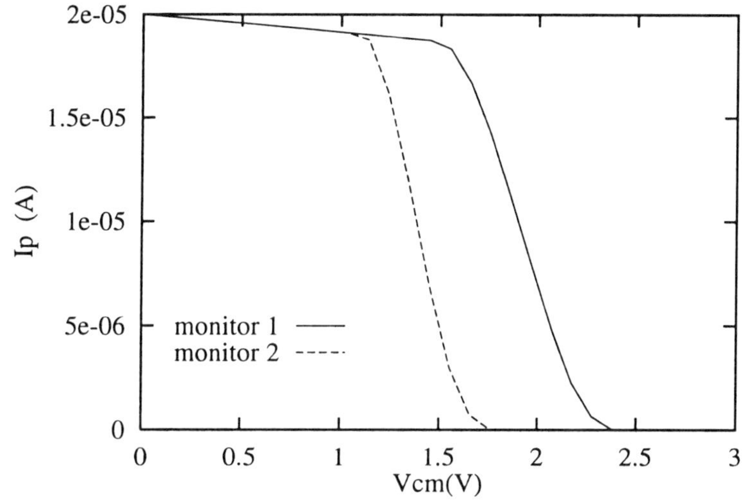

(a)

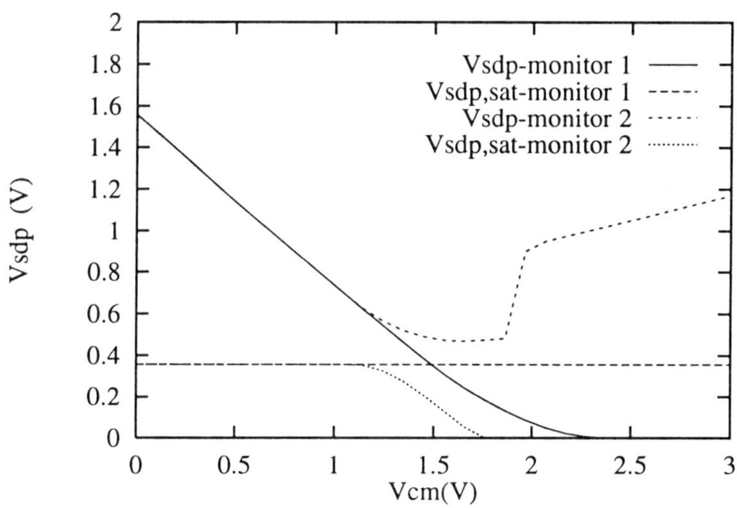

(b)

Figure 4.9: Simulation results of monitor circuits: (a) drain current I_p, (b) V_{SDp} and $V_{SDp,sat}$ of M_p as a function of V_{CM}.

Chapter 5

Constant-g_m Input Stages, $K_n \neq K_p$

5.1 Constant-g_m Input Stages

In this chapter, complete constant-g_m input stages will be finally developed. Figure 5.1 shows the input stage which consists of the bias circuit 2(Figure 4.3) and the monitor circuit 1(Figure 4.7). Based on design concepts we used, this circuit should work as a constant-g_m input stage; however, the circuit as shown in Figure 5.1 has a couple of problems that needs to be addressed.

The first task is to find the relationships between I_{ref}, I_c, and I_d. The bias circuit should keep g_{mT} constant for any value of the common mode input voltage V_{CM}, *i.e.*,

$$\begin{aligned} g_{mT} &= \sqrt{2I_cK_p} + \sqrt{2I_dK_n} \\ &= \sqrt{2I_pK_p} + \sqrt{2I_nK_n}. \end{aligned} \tag{5.1}$$

When V_{CM} is small enough such that M_r, and M_q, and M_p are operating as a current mirror, the p-channel differential pair receives I_{ref} while the n-channel pair is turned off. Then we have

$$g_{mT} = \sqrt{2I_{ref}K_p} \tag{5.2}$$

and so we see that in order to have a constant-g_m characteristics, we must satisfy,

$$\sqrt{I_cK_p} + \sqrt{I_dK_n} = \sqrt{I_{ref}K_p}. \tag{5.3}$$

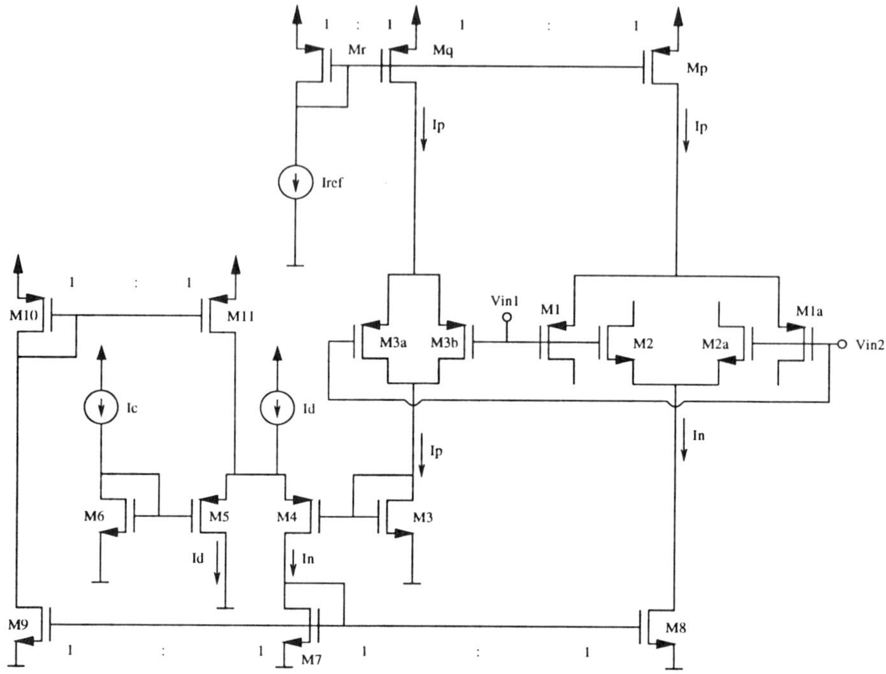

Figure 5.1: Constant-g_m input stage using monitor circuit 1 and the bias circuit 2.

If we assume that $I_{ref} = I_c X$ where X is larger than one, (5.3) can be rewritten as

$$\sqrt{K_p I_c}(\sqrt{X} - 1) = \sqrt{K_n I_d} \tag{5.4}$$

and if we choose $X = 4$, (5.4) becomes

$$K_p I_c = K_n I_d. \tag{5.5}$$

This is one of the conditions that must be implemented; the circuit realization for the relation given in (5.5) will be given later in this chapter. Until then we will set $I_c = I_d = I_{ref}/4$ assuming that $K_n = K_p$ can be realized. Since we know μ_n and μ_p, we can temporarily meet (5.5) by choosing the appropriate transistor sizes; this is done to separate this condition from another condition pertaining to operation in subthreshold(weak inversion) which is discussed next.

5.2 Weak Inversion Region Operation

So far all circuit designs were carried out assuming the square-law characteristics of a MOS transistor, that is, we assumed that the transistors are operating in the saturation region in the strong inversion region. In a square-law equation as given in (2.1), the drain current is assumed to be zero if the gate to source voltage , V_{GS}, of a transistor is less than its threshold voltage, V_T. However, in reality when $V_{GS} < V_T$ the transistor simply leaves the strong inversion and enters the weak inversion region in which the relationship between the current and the voltages is an exponential one and is given by[7]

$$I_D = 2nU_T^2 Kexp\left(\frac{(V_G - V_T)/n - V_S}{U_T}\right) \tag{5.6}$$

where $U_T = kT/q$ and n is a constant between 1 and 2. The equation in (5.6) assumes that $V_{DS} > 3KT/q$ so that the transistor is operating in saturation region. Note that in reality there is a small region called moderate inversion region between the weak and the strong inversion regions. Since analog circuits are usually designed with the assumption that transistors are either in deep strong inversion or in deep weak inversion region, little attention has been given to moderate inversion and therefore simulation tools such as SPICE or APLAC have not incorporated proper models for moderate inversion.

Figure 5.2 shows I-V curves of an n-channel MOS transistor simulated using two different device models. LEVEL 1 model more or less follows the square-law characteristics which has a sharp turn on at $V_{gate} \approx V_T$(about 0.9V); this can be better seen on the curve that is magnified by 100. Smoother I-V characteristics are seen on a curve that was generated using the BSIM model. Note that there is a small current flow for V_{gate} well below V_T. In other words, when I_d becomes very small, V_{GS} becomes less than V_T. The effect of the weak inversion operation is discussed next and a solution to this problem will be provided.

Referring to Figure 5.1, the purpose of the subcircuit consisting of $M_3 - M_6$ was to maintain the relationship given in (4.8) so that g_{mT} is constant. However, to reach (4.8), we made an assumption that all the transistors are operating in the saturation region and V_{T3} and V_{T6} cancel each other in (4.6). When V_{CM} becomes close to V_{DD}, I_p that flows into M_3 becomes very small and the assumption is only half true, *i.e.*, it is still operating in saturation region, but it is no longer in the strong inversion. If I_p becomes zero, according to (4.6) V_{GS3} is represented by V_{T3} which is a constant; However, it was just shown that when the transistor is in weak inversion, V_{GS} decreases with I_d according to (5.6). Since the sum of V_{SG4} and V_{GS3}

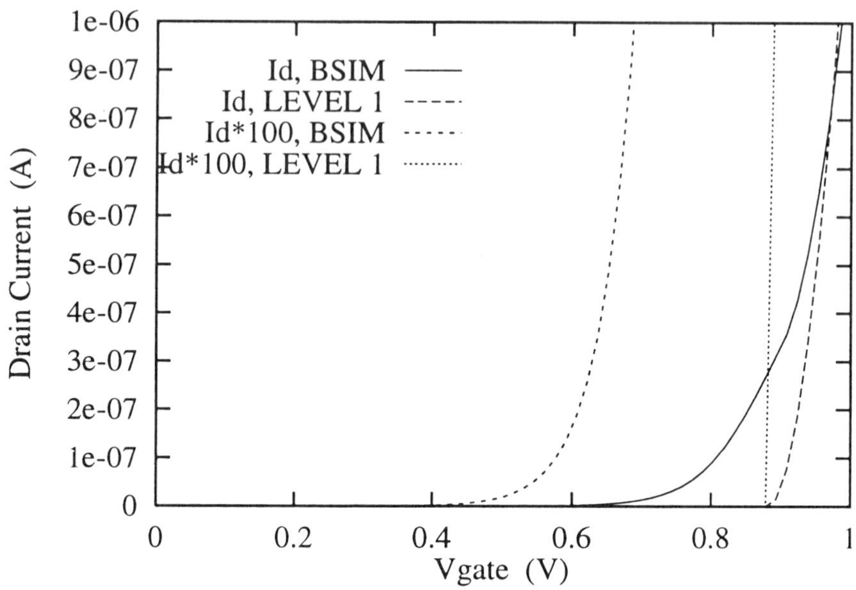

Figure 5.2: I-V curves of an n-channel transistor simulated with BSIM and LEVEL 1 models

is guaranteed to be constant, when V_{GS3} becomes less than its intended value(V_{T3}), V_{SG4} becomes larger than its maximum intended value. This of course makes I_n and hence g_{mn} larger than what they should be when I_p is very small and g_{mT} is no longer constant. Figure 5.3 shows V_{GS} of M_3 as a function of V_{CM}. It can be seen that as V_{CM} is increased, V_{GS3} is reduced as a result of the decrease in I_p. When V_{CM} is about 2.2V, V_{GS3} equals V_T. When V_{CM} is increased further, V_{GS3} does not remain at V_T but it keeps decreasing. The effect of this is shown in Figure 5.4; I_n and g_{mn} for $V_{CM} > 2.5V$(only the n-channel differential pair contributes to g_{mT}) are significantly larger than those for $V_{CM} < 0.5$(only the p-channel pair contributes to g_{mT}). Note that in this simulation, V_{GS3} converged to about 0.5V for large V_{CM}; however, in reality this value is unpredictable and therefore the maximum value of g_{mn} is also unpredictable. That is, this problem cannot be corrected by simply optimizing the transistor sizes, and it requires an additional circuit technique to be used.

One way to overcome the problem caused by the weak inversion region

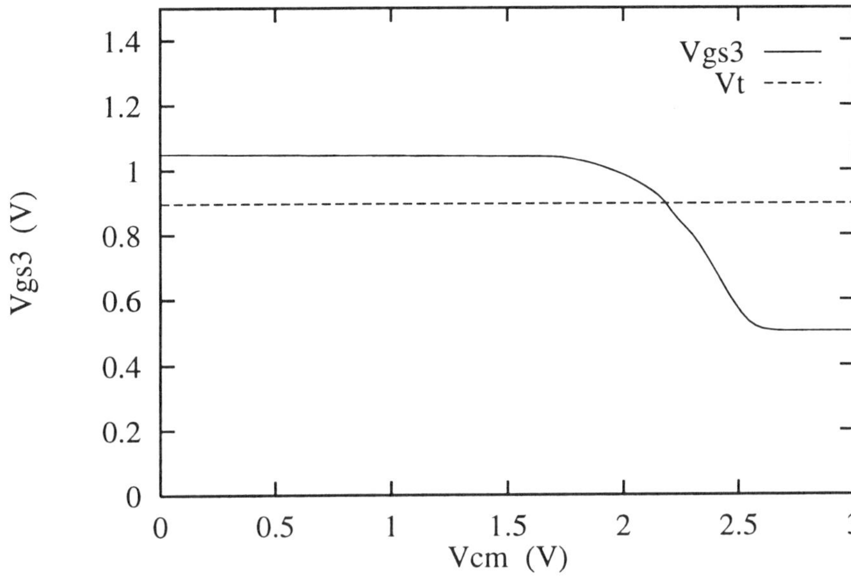

Figure 5.3: V_{GS3} of the input stage, showing the effect of weak inversion as a function of V_{CM}

operation of M_3 is to hard limit the value of I_n just as I_p is limited to I_{ref}. In the circuit as given in Figure 5.1 there exists a current path for I_n consisting of $M_4 - M_7 - M_9 - M_{10} - M_{11}$ back to M_4. There is no mechanism that limits the value of I_n except by putting transistors into triode region when I_n is too large. Obviously, it will be difficult to design this circuit so that some of the transistors leave saturation region as soon as M_3 enters the weak inversion region. Instead, the approach will be to break the current path for I_n and somehow limit it to I_{nmax} the maximum value of I_n. The circuit shown in Figure 5.5 is a modified version of the bias circuit 2 which is shown in Figure 4.3. The source terminals of M_4 and M_5 are separated, and the voltages at these nodes are maintained to equal each other by the circuit consisting of M_{10}, M_{11}, M_{12}, and M_{13}. Note that these devices can be recognized as part of a current conveyer[55]. Because of the current mirror $M_{12} - M_{13}$, the drain currents in M_{10} and M_{11} are equal to each other. Then their gate to source voltages should be the same; since their gate terminals are connected, the source terminals of M_{10} and M_{11} or

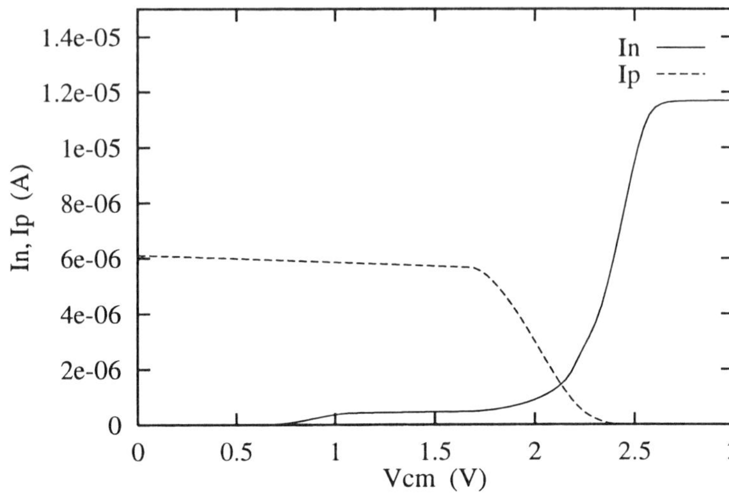

(a)

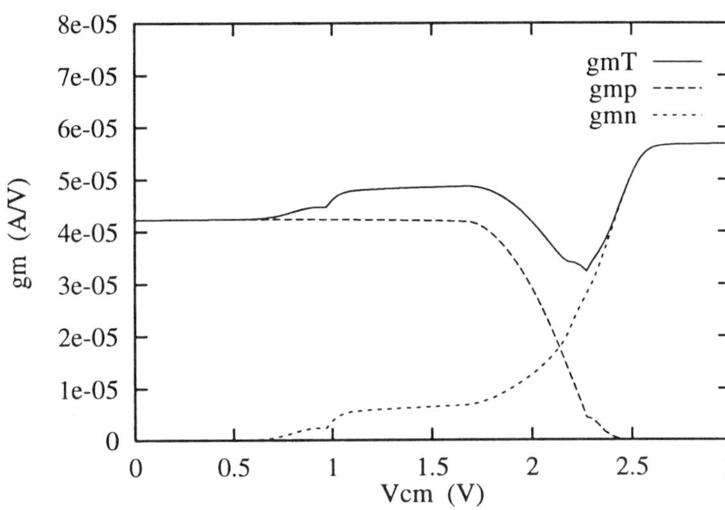

(b)

Figure 5.4: Simulation results of the input stage showing the effect of weak inversion: (a) Differential pair currents (b) Differential pair transconductance.

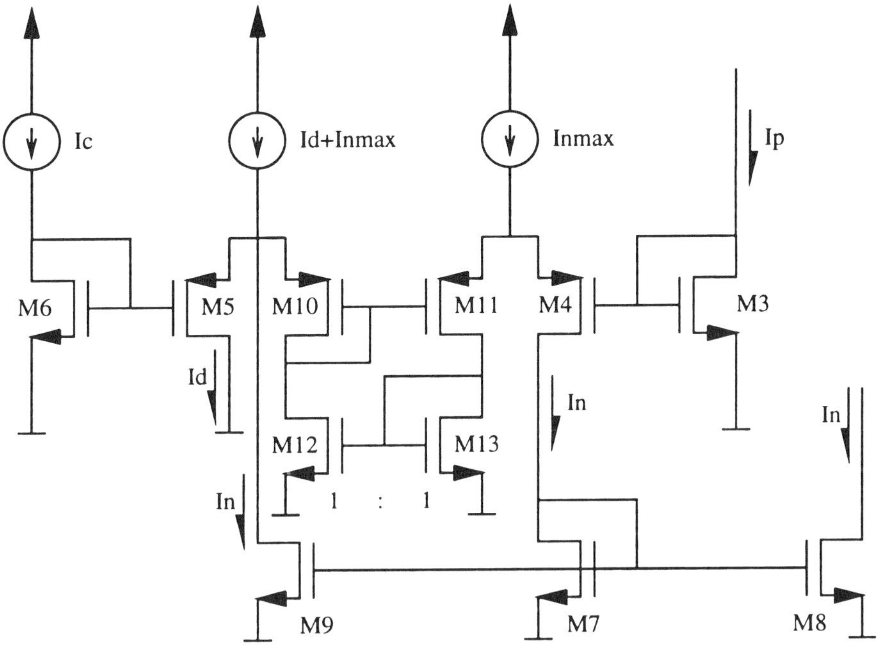

Figure 5.5: Modified version of the bias circuit 2. This implementation overcomes the problem caused by M_3 going into the weak inversion region

similarly the source terminals of M_4 and M_5 are at the same potential as desired. A constant current source I_{nmax} supplies currents to the sources of M_4 and M_{11}, and no matter how small V_{GS3} becomes, I_n is limited to I_{nmax}. A constant current source $I_d + I_{nmax}$ supplies currents to the sources of M_5 and M_{10}; adding M_9 to sink I_n from this node forces the drain current in M_5 to equal I_d. Thus, as they were in bias circuit 2, M_5 and M_6 are biased with I_d and I_c, respectively.

Before we place the modified bias circuit of Figure 5.5 into the input stage, there is one more aspect that needs to be taken care of. In the previous section, it was pointed out that in order to have the same g_{mT} when only the p-channel differential pair is operating and when both pairs are working, the relationship given in (5.5) has to be satisfied. Since the maximum value of I_n is determined to be I_{nmax} thanks to the circuit shown in Figure 5.5, g_{mn} that results from I_{nmax} must now be made equal to (5.3) also. Because I_{ref} is the maximum value for I_p, let us rename I_{ref} as I_{pmax}; then when only the p-channel differential pair is operating, we have

$$g_{mT} = g_{mp} = \sqrt{2 I_{pmax} K_p}. \tag{5.7}$$

This will be the case for V_{CM} small enough such that $I_p = I_{pmax}$. When V_{CM} is increased so that I_p becomes less than I_{pmax} and the n-channel pair is also contributing to g_{mT}, we have

$$\begin{aligned} g_{mT} &= g_{mn} + g_{mp} \\ &= \sqrt{2K_n I_d} + \sqrt{2K_p I_c}. \end{aligned} \tag{5.8}$$

When V_{CM} is large enough such that I_p becomes zero and only the n-channel pair is operating, we have

$$g_{mT} = g_{mn} = \sqrt{2I_{nmax}K_n}. \tag{5.9}$$

In order to have a constant-g_m input stage, (5.7), (5.8), and (5.9) must be made the same. First , to make (5.7) and (5.9) equal, we must have

$$I_{pmax}K_p = I_{nmax}K_n. \tag{5.10}$$

Notice that if we chose $I_c = I_{pmax}/4$ and $I_d = I_{nmax}/4$, (5.5) is satisfied and (5.8) will also be the same as (5.7) and (5.9). This makes perfect sense since the condition given in (5.5) was sufficient to achieve (5.3). The circuit that realizes the condition of (5.9) is shown in Figure 5.6. The circuit is supplied with a constant bias current $(1 + a)I_{nmax}$. there are two current mirrors which have the current ratio of 1:a. The main part of the circuit consists of transistors M_{1B}, M_{2B}, M_{3B}, and M_{4B} which are used in the exact same configuration as the constant-g_m bias circuit 2, and we have

$$V_{SG3B} + V_{GS2B} = V_{SG4B} + V_{GS1B}. \tag{5.11}$$

Since all the currents are constant, these is no need to be concerned with transistors going into the weak inversion region. Then assuming the square-law characteristics, (5.11) becomes

$$\sqrt{\frac{I_{pmax}}{K_n}} + \sqrt{\frac{aI_{nmax}}{K_p}} = \sqrt{\frac{I_{nmax}}{K_p}} + \sqrt{\frac{aI_{pmax}}{K_n}} \tag{5.12}$$

which, after some algebra, becomes the same as (5.10). In order to demonstrate the effectiveness of this circuit, I_{nmax} and I_{pmax} are used as the bias currents for n- and p-channel differential pairs, respectively, and g_{mn} and g_{mp} are compared. In the simulation, the gates of n- and p-channel pairs are connected to V_{DD} and V_{SS}, respectively, and the width of the transistors are varied so that the ratio of K_n to K_p is changed. As long as (5.10) is realized by this circuit, g_{mn} and g_{mp} of the differential pairs should be equal regardless of the value used for the ratio K_n/K_p. Figure 5.7 shows the

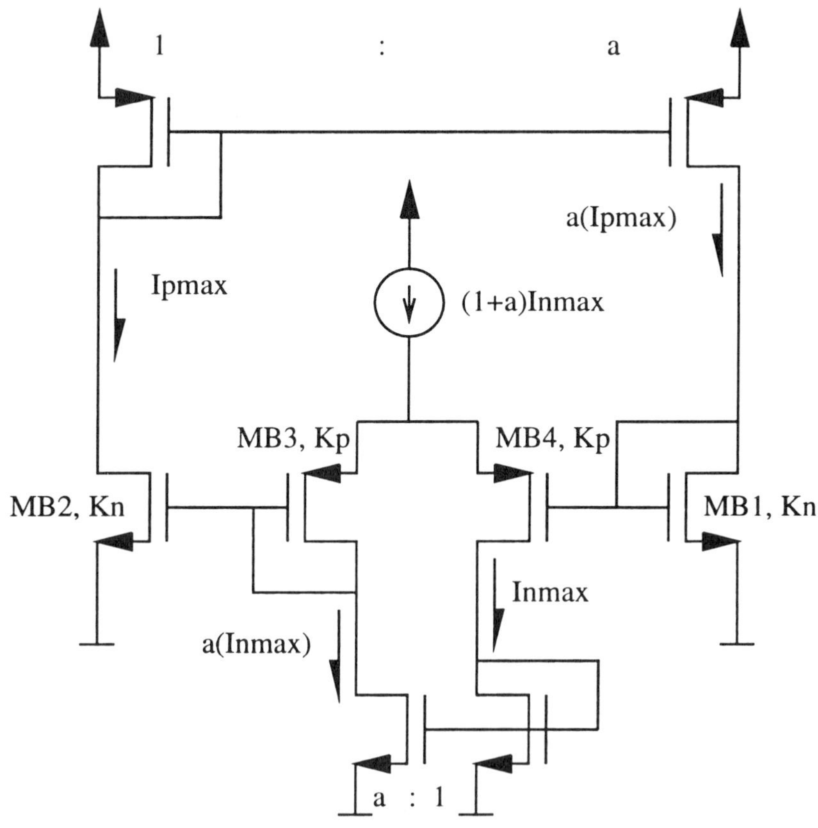

Figure 5.6: A CMOS circuit that satisfies the condition $I_{pmax}K_p = I_{nmax}K_n$.

simulation results of the circuit as the width of the n-channel transistors, W_n, is varied. The original width of the n-channel transistors is W_{n0} and W_n/W_{n0} is changed from 0.7 to 1.3. In Figure 5.7(a), I_{nmax} is constant while I_{pmax} increases with increasing W_n. Since the bias current feeding the source terminals of M_{B3} and M_{B4} is constant, I_{nmax} should be constant whereas I_{pmax} must increase to compensate for the reduction in K_p/K_n to keep g_{mn} and g_{mp} equal; this is shown in Figure 5.7(b). Figure 5.8 shows the simulation results of the same circuit with W_p/W_{p0} being varied. W_p is the transistor width of the p-channel transistors and W_{p0} is the initial width. It can be seen in Figure 5.8(a) that I_{pmax} increases with increasing W_p as expected, and the closeness of g_{mn} and g_{mp} is shown in Figure 5.8(b).

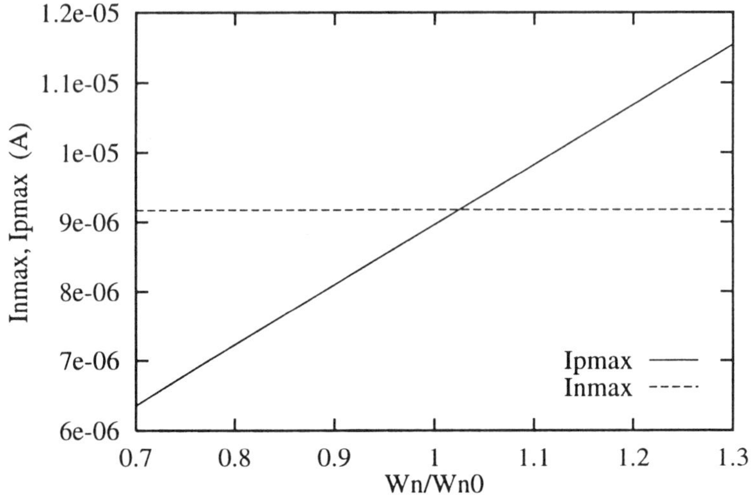

(a)

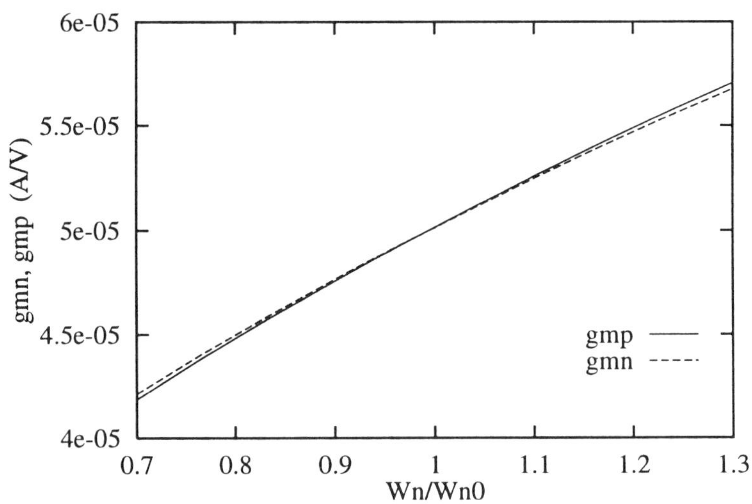

(b)

Figure 5.7: Simulation results of the circuit, which maintains $I_{pmax}K_p = I_{nmax}K_n$, as a function of W_n/W_{n0}.

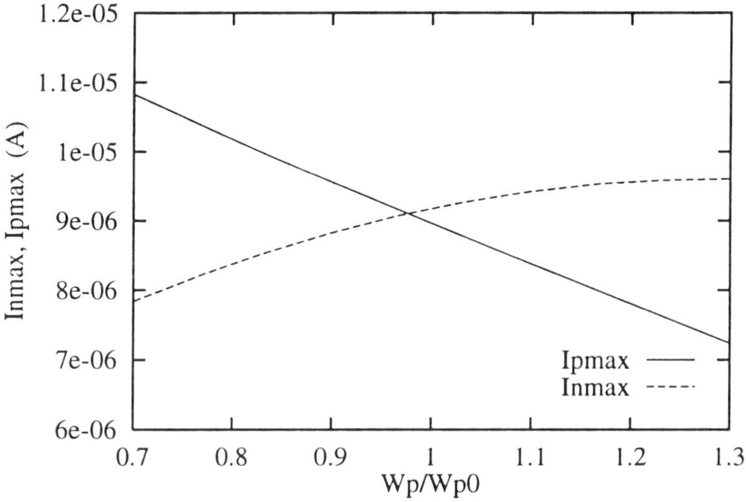

(a)

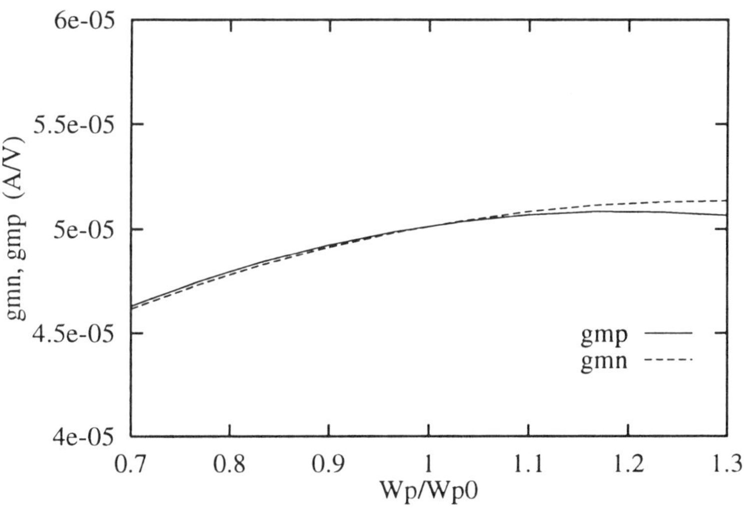

(b)

Figure 5.8: Simulation results of the circuit, which maintains $I_{pmax}K_p = I_{nmax}K_n$, as a function of W_p/W_{p0}.

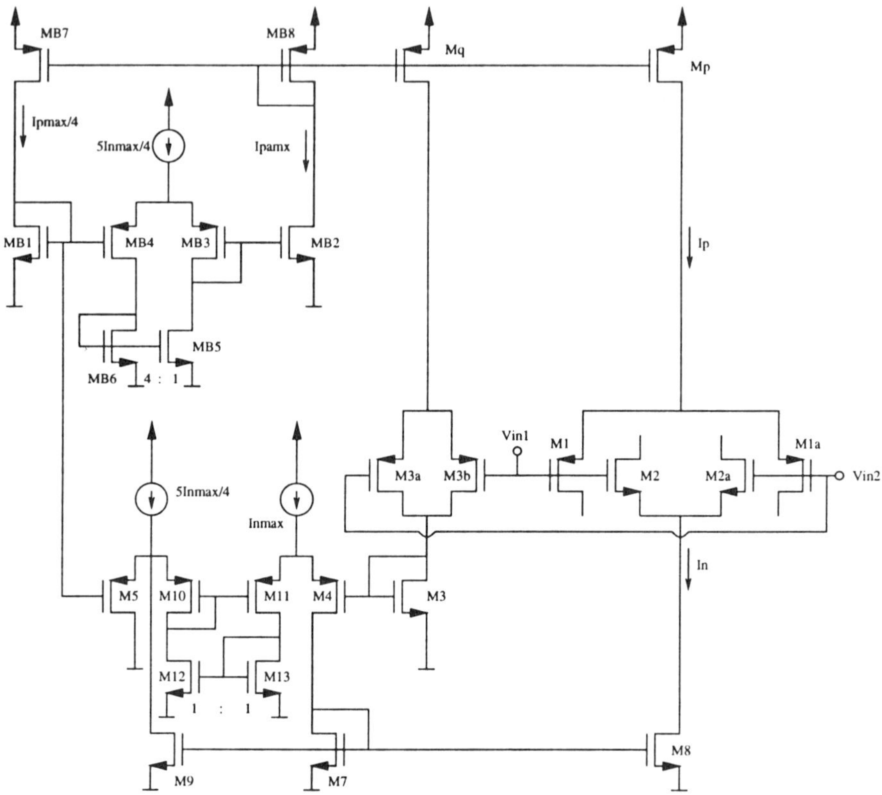

Figure 5.9: Constant-g_m input stage 1.

5.3 Two New Constant-g_m Input Stages

At this point, we have all the necessary subcircuits to construct a constant-g_m input stage. Actually since there are two different types of monitor circuits which seem to work equally well, two input stages can be built and are discussed next.

Constant-g_m Input Stage 1 is shown in Figure 5.9. This circuit uses the first monitor circuit(monitor 1). Since M_6(not shown) and M_{B1} are both biased with $I_{pmax}/4$, their gate voltages are the same. Since the purpose of M_6 is to bias the gate terminal of M_5, it can be eliminated simply by connecting the gate of M_5 to the gate of M_{B1}. Simulation results of this input stage are shown in Figure 5.10. The variation in g_{mT} as V_{CM} is swept from 0 to 3V is less than 10%; although this is not perfectly constant, it is much better than the 100% deviation which results from input stages without any novel biasing schemes. Furthermore, the circuit shown in Figure 5.9 does not require matching between n- and p-channel transistors.

Constant-g_m Input Stage 2 is shown in Figure 5.11. This circuit is exactly the same as the constant-g_m input stage 1 except that monitor circuit 2 is used instead of monitor circuit 1. The bias voltage, V_b, which is the gate voltage of M_b is set to 1.7V. The simulation results are shown in Figure 5.12; the curves are similar to those obtained for the constant-g_m circuit 1 except that they are slightly shifted to the left. This is due to the reference voltage V_b which determines the point at which g_{mn} and g_{mp} are the same.

In the g_{mT} plot of both circuits, most of the deviations are caused by the bumps that show up at values of V_{CM} such that one of the differential pairs is conducting a very small amount of current. There are two apparent reasons for these bumps. First, when the drain currents of the differential pairs are small, transistors go into weak inversion region and the transistor model used for the simulation is not accurate enough. That is, we are hoping that the bumps will get smaller when the actual g_{mT} is measured from a fabricated chip. Second, when one of the input pairs is operating in the weak inversion region, g_m of that pair is no longer proportional to the square root of the current. Even though the contribution of that pair to the total g_m is much less than that of the other pair which is conducting a lot of current, it may cause a several percent deviation in g_{mT}. This point is further investigated in the next section.

5.4 Effects of Operation in Subthreshold

It was shown in the previous section that g_{mT} obtained from the constant-g_m input stages is not perfect, partially due to the fact that some transistors go into the weak inversion region. Figure 5.13 shows the gate to source voltages and the threshold voltages of the input transistors as a function of V_{CM}. In Figure 5.13(a), it can be seen that V_{GSn} and V_{SGp} intersects V_{Tn} and $|V_{Tp}|$ at V_{CM} equal to $1.65V$ and $2.3V$, respectively. These points indicate the value of V_{CM} at which the respective transistors go into weak inversion region. Note that these points are at the locations where the bumps appear in g_{mT} plot of Figure 5.10(b). Thus, it leads us to believe that the weak inversion region operation of the transistors causes the major portion of the deviation in g_{mT}. Exactly the same conclusion can be drawn by comparing the plots in Figure 5.13(b) and Figure 5.12(b).

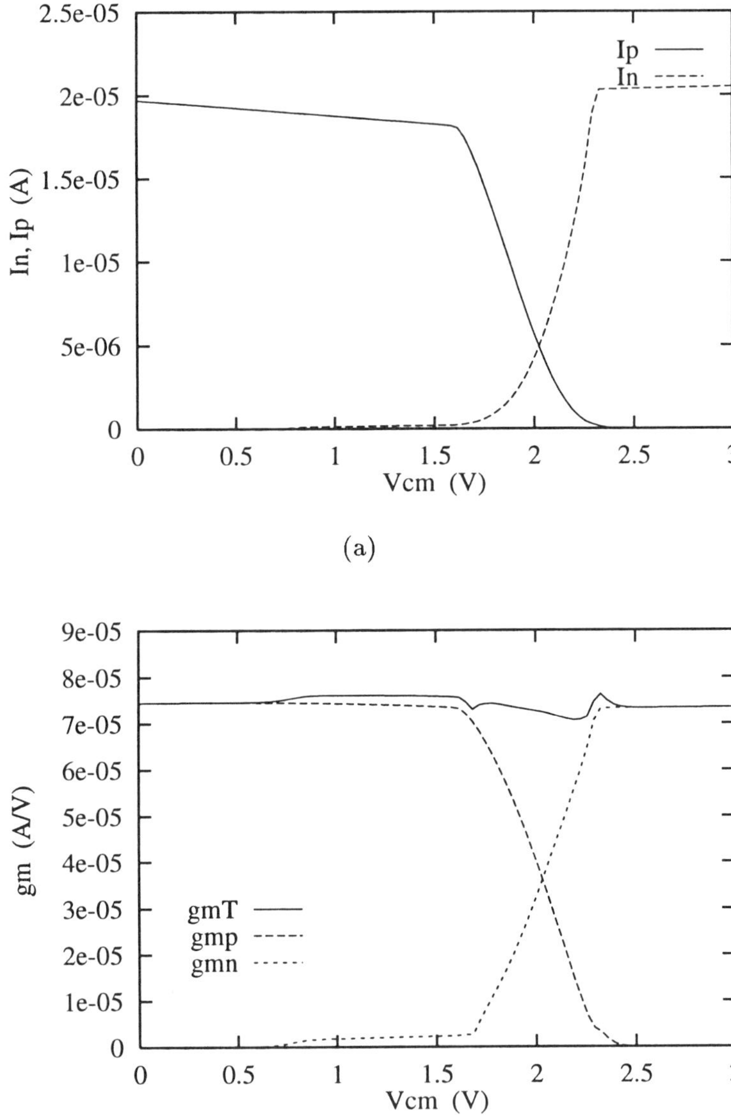

(b)

Figure 5.10: Simulation results of the constant-g_m input stage 1.

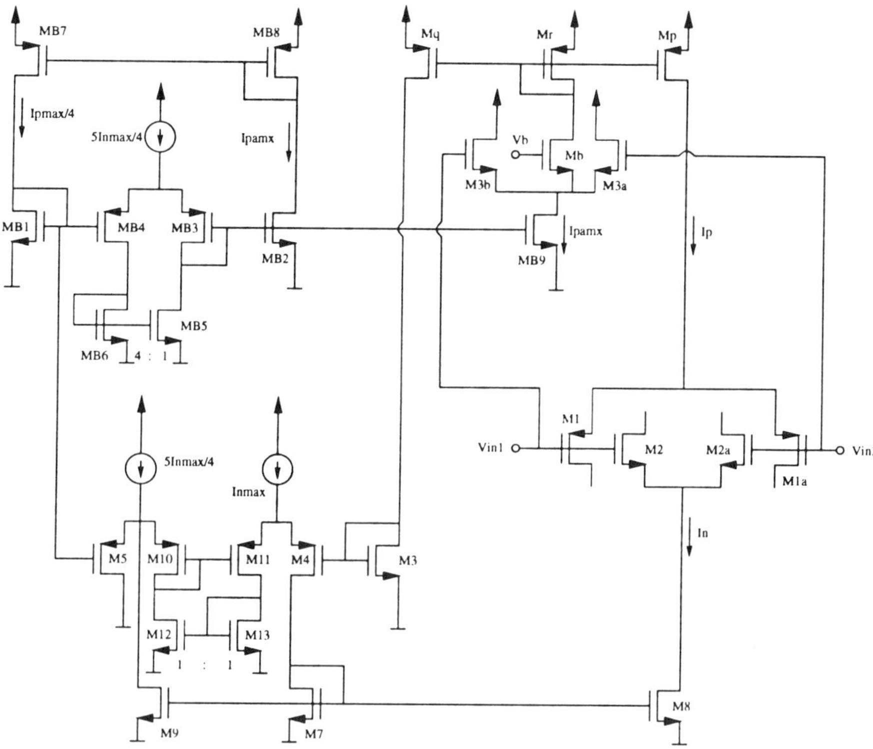

Figure 5.11: Constant-g_m input stage 2.

The constant-g_m input stage will be analyzed assuming ideal conditions except: 1) for very small I_n, $M_2(M_{2a})$ and/or M_4 will be in weak inversion region, and 2) for very small I_p, $M_1(M_{1a})$ and/or M_3 will be in weak inversion region. Using (5.6), and by setting $n = 1$ for simplicity, we can find that a transistor operating in the weak inversion region has

$$g_m = I/U_T \tag{5.13}$$

and

$$V_{GS} = U_T ln \frac{I}{2U_T^2 K} + V_T. \tag{5.14}$$

As V_{CM} is swept from rail-to-rail, the input differential pairs go through three operating regions. When V_{CM} is sufficiently close to the negative rail, I_n is small, then

$$g_{mT} = \sqrt{2K_p I_p} + \frac{I_n}{2U_T}. \tag{5.15}$$

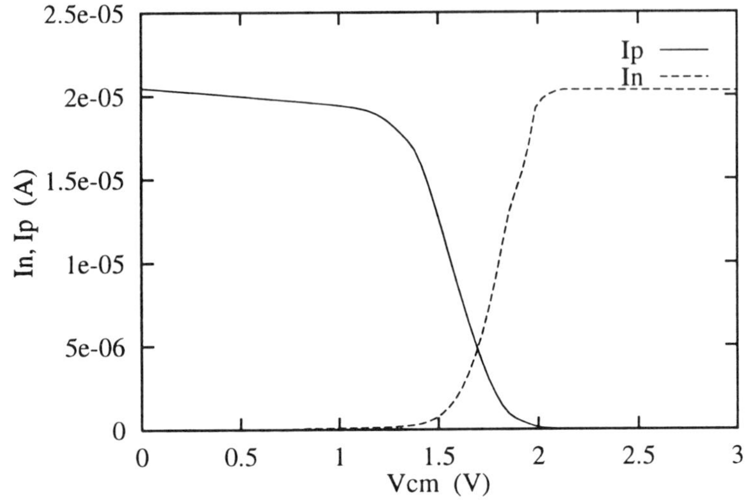

(a)

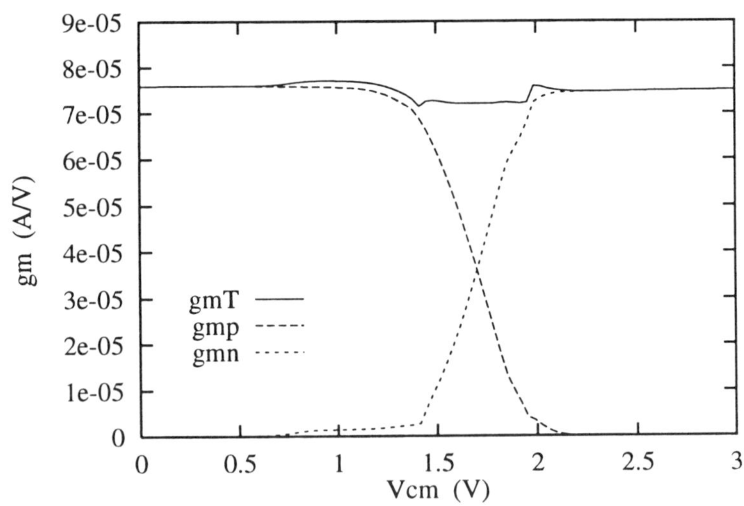

(b)

Figure 5.12: Simulation results of the constant-g_m input stage 2.

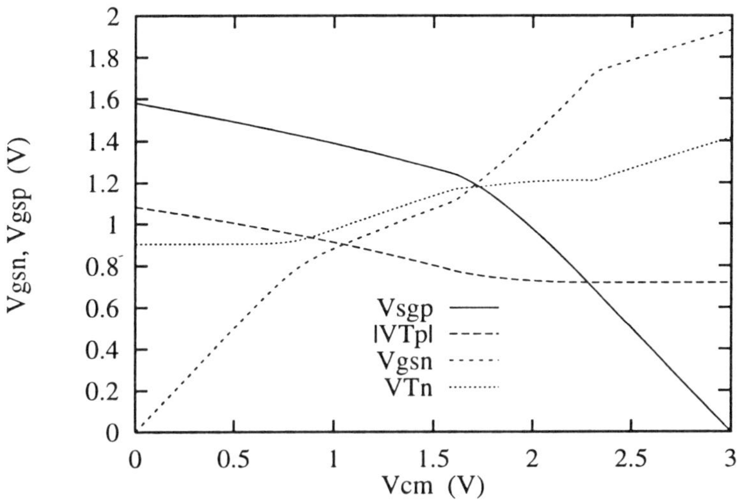

(a)

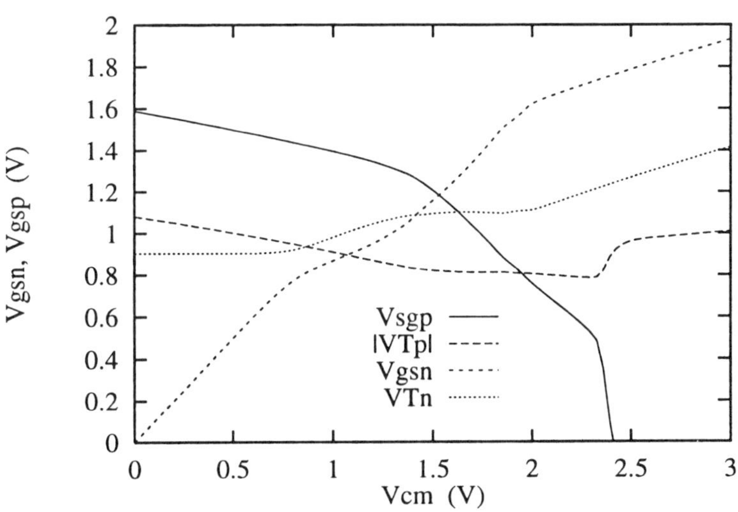

(b)

Figure 5.13: Gate to source voltages and the threshold voltages of the input transistors of: (a) constant-g_m input stage 1, and (b) 2.

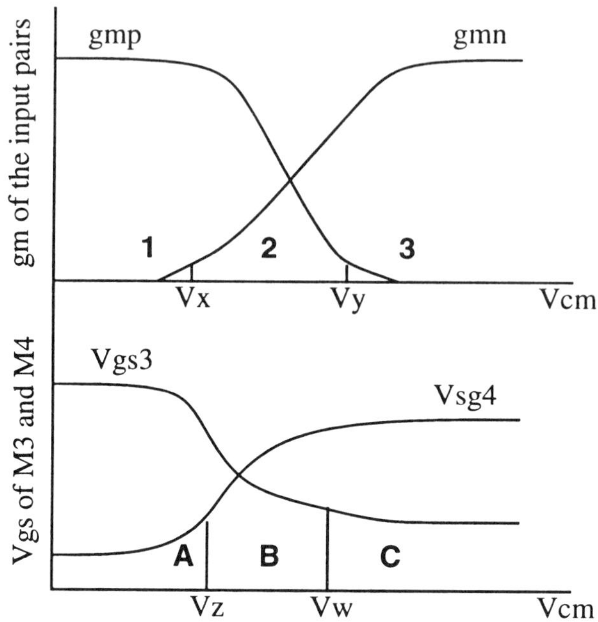

Figure 5.14: Different operating regions for input differential pairs and $M_3 - M_4$ pair.

We will denote this region as **region 1** as shown in Figure 5.14 for $V_{CM} < V_X$. When V_{CM} is sufficiently away from both rails, both pairs are operating in the strong inversion region, and g_{mT} is given by

$$g_{mT} = \sqrt{2K_pI_p} + \sqrt{2K_nI_n} \tag{5.16}$$

which is the nominal expression for g_{mT} and let us denote this region as **region 2**. When V_{CM} is close to the positive rail, I_p is small, and

$$g_{mT} = \frac{I_p}{2U_T} + \sqrt{2K_nI_n} \tag{5.17}$$

which describes the operation of the differential pairs in **region 3**($V_{CM} > V_Y$ in Figure 5.14). Although the sum of V_{SG4} and V_{GS3} is constant, the combination of M_3 and M_4 goes through three operating regions as the differential pairs; however, it is not necessary that the differential pairs and the $M_3 - M_4$ combination change their operating region simultaneously, that is $V_X = V_Z$ and $V_Y = V_W$ may not be true. Note that V_Z and V_W are the values of V_{CM} for which $V_{SG4} = |V_{Tp}|$ and $V_{GS3} = V_{Tn}$, respectively, take place. In fact, since V_X is for the n-channel transistor and V_Z is for the p-channel transistor, and although these values are naturally close to each other, it is almost impossible to intentionally match them. Thus we must separately define three regions for the $M_3 - M_4$ pair. In **region A**, where $V_{CM} < V_Z$, I_n is small and M_4 is in the weak inversion region. We then have

$$\sqrt{K_p I_{pmax}/2} + \sqrt{K_n I_{nmax}/2} = \sqrt{2K_p I_p} + \sqrt{2K_n K_p}(U_T ln \frac{I_n}{2U_T^2 K_p}). \tag{5.18}$$

In **region B**, both transistors are operating in the strong inversion region and,

$$\sqrt{K_p I_{pmax}/2} + \sqrt{K_n I_{nmax}/2} = \sqrt{2K_p I_p} + \sqrt{2K_n I_n} \tag{5.19}$$

which describes the nominal operation of the $M_3 - M_4$ pair. In **region C**, where $V_{CM} > V_W$, I_p is small and M_3 is in weak inversion region; this results in

$$\sqrt{K_p I_{pmax}/2} + \sqrt{K_n I_{nmax}/2} = \sqrt{2K_n K_p}(U_T ln \frac{I_p}{2U_T^2 K_n}) + \sqrt{2K_n I_n} \tag{5.20}$$

The different combinations of the differential pairs operating regions and the $M_3 - M_4$ pair operating regions constitute the operating regions of the input stage as a whole. For each combination, there are two equations, one of (5.15), (5.16), or (5.17) and one of (5.18), (5.19), or (5.20), and two unknows, I_n and I_p. Table 5.1 shows the possible operating regions and the conditions for their existence. Even though, it is theoretically possible to determine I_n and I_p for each operating region and hence find g_{mT}, it is not worthwhile analyzing the regions 1B, 2A, 2C, and 3B for reasons given below. Using the simple drain current equations it is impossible to precisely determine the locations of V_X, V_Y, V_Z, and V_W , and this makes it difficult to decide whether or not the equations that are used are valid. Also, because these regions are so narrow, the knowledge of the regions 1A, 2B, and 3C which are right next to them should provide us with enough information to roughly estimate g_{mT} in these regions.

We will now concentrate on regions 1A, 2B, and 3C. For simplicity, let us assume that the n- and p-channel transistors are matched, *i.e.*, $K_n = K_p = K$ and $V_{Tn} = |V_{Tp}| = V_T$, then regions 1A and 3C are basically the same and they will be compared with region 2B which is the region that possesses the nominal characteristics. In the calculations, the constants are set as: $I_{pmax} = I_{nmax} = 20\mu A$, $V_T = 0.8V$, $U_T = 25mV$, and $K = 100, 200,$ or $300\mu A/V^2$. The percentage error of g_{mT} is calculated as the difference of

Table 5.1: Possible operating regions of the input stage.

Region	$M_1 - M_2$	$M_3 - M_4$	Conditions	Equations
1A	1	A	None	(5.15) & (5.18)
1B	1	B	$V_X > V_Z$	(5.19) & (5.15)
2A	2	A	$V_Z > V_X$	(5.16) & (5.18)
2B	2	B	None	(5.16) & (5.19)
2C	2	C	$V_Y > V_W$	(5.16) & (5.20)
3B	3	B	$V_W > V_Y$	(5.17) & (5.19)
3C	3	C	None	(5.17) & (5.20)

g_{mT} in region 1A and 2B divided by g_{mT} in 2B. Calculations were performed for three different values of K, and the results are shown in Figure 5.15(a). Note that the x-axis in the figure corresponds to I_n and I_p for regions 1A and 3C, respectively. Figure 5.15(b) shows the gate to source voltages of the transistors which are conducting small currents. Since it is assumed that in regions 1A and 3C, transistors with small currents are in weak inversion region, V_{GS}(or V_{SG}) of those transistors should be less than $V_T = 0.8V$. Thus, for the case with $K = 100\mu A/V^2$, using Figure 5.15, we see that I must be less than about $0.1\mu A$ and the curves for $K = 100\mu A/V^2$ in Figure 5.15 can be trusted up to that point. Using this method to determine the valid range in Figure 5.15, it seems that larger K causes the peak error to be larger. However, we know that real transistors with V_{GS} close to V_T may still be in the moderate inversion region, and the drain currents may be significantly smaller than values determined above for weak inversion operations. For instance, if we look at Figure 5.15(a) for only the part with $I < 0.1\mu A$, the error is larger for smaller K, opposite from the conclusion drawn above. While the analysis given here does not provide us with the precise value of the error caused by the weak inversion operation nor does it enable us to choose the optimal value of K, because the current and the transconductance of a transistor varies continuously between its operating regions, we know that the error will not be significantly larger than the predictions provided here.

5.5 Other Nonideal Effects

We have so far assumed that if all the transistors are in saturation in the strong inversion region, g_{mT} will be constant as a function V_{CM}. Using the square-law relationship of (2.1), the total transconductance of the differential pairs is given by (4.1) which was maintained constant by the

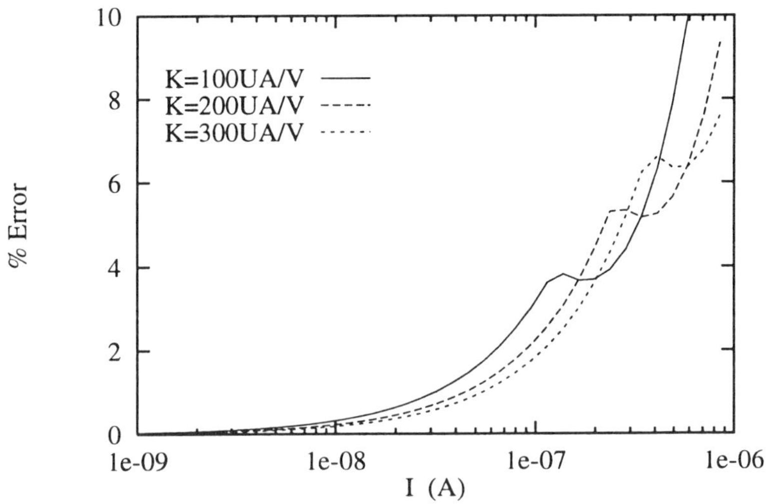

(a)

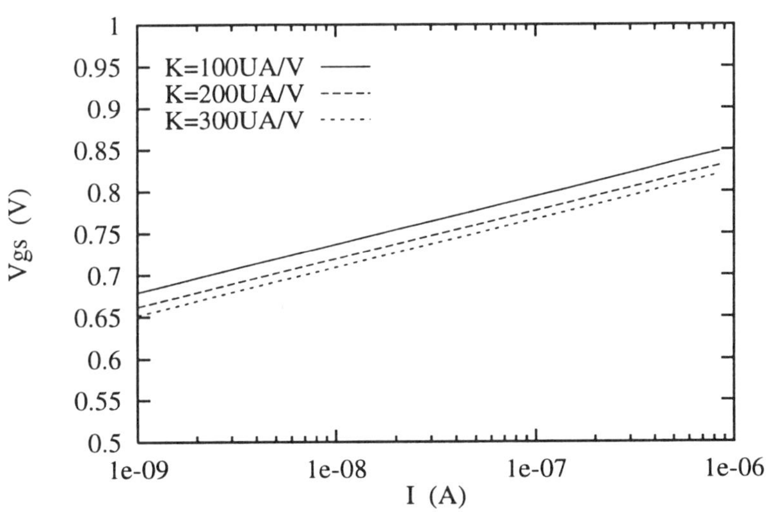

(b)

Figure 5.15: Calculated percentage error in g_{mT} caused by the weak inversion operation of the transistors in the input stage.

constant-g_m circuit having the characteristics described by (4.8). However, due to second order effects, transistors do not exhibit perfect square-law characteristics, thus, g_{mT} is not expected to be perfectly constant. In the following, the effect of mobility degradation and the body effect on the performance of both the differential pairs and the constant-g_m bias circuit are considered; other parts of the input stage, such as the current mirrors, are assumed to be ideal. Channel length modulation will not be considered since it will introduce the dependence of the circuit performance on the drain voltages of the transistors and this will require knowledge of the circuits connected to the drain terminals. Moreover, it is well known that the effect of the channel length modulation can be reduced by choosing a larger gate length. This observation should lead to a design procedure to reduce the error in g_{mT} caused by the channel length modulation.

Mobility degradation causes the transconductance parameter, K, of a transistor to decrease with increasing V_{GS}. The drain current equation of a transistor taking into account this second order effect is given by[56]

$$I = \frac{K}{1 + \theta(V_{GS} - V_T)}(V_{GS} - V_T)^2 \tag{5.21}$$

where θ is a process dependent mobility degradation constant which is usually between 0.01 and $0.25V^{-1}$. Using (5.21), the gate to source voltage and the transconductance can be written approximately as

$$V_{GS} = \sqrt{\frac{I}{K}} + V_T + \frac{\theta I}{2K} \tag{5.22}$$

and

$$g_m = 2\sqrt{KI} - 2\theta I, \tag{5.23}$$

respectively. Using the new expression for V_{GS}, it can be shown that the constant-g_m bias circuit 2 shown in Figure 4.3 has the following relation according to (4.5).

$$\sqrt{\frac{I_{pmax}K_p}{2}} + \sqrt{\frac{I_{nmax}K_n}{2}} + \theta_n I_{pmax}\sqrt{\frac{K_p}{32K_n}} + \theta_p I_{nmax}\sqrt{\frac{K_n}{32K_p}}$$
$$= \sqrt{2I_nK_p} + \sqrt{2I_pK_n} + \theta_n I_p\sqrt{\frac{K_p}{2K_n}} + \theta_p I_n\sqrt{\frac{K_n}{2K_p}}. \tag{5.24}$$

Note that $I_c = I_{pmax}/4$ and $I_d = I_{nmax}/4$ were used, and (5.24) is the expression that is being maintained constant; let us call it G_m. Using (5.23), the total transconductance of the differential pairs is now given by

$$g_{mT} = \sqrt{2K_nI_n} + \sqrt{2K_pI_p} - \theta_p I_p - \theta_n I_n. \tag{5.25}$$

In order to gain more insight into the significance of above expressions, let us assume that $K_n = K_p = K$ and $\theta_n = \theta_p = \theta$, then (5.24) and (5.25) become

$$g_{mT} = \sqrt{2K}(\sqrt{I_p} + \sqrt{I_n}) - \theta(I_n + I_p) \quad (5.26)$$

$$G_m = \sqrt{2K}(\sqrt{I_p} + \sqrt{I_n}) + \frac{\sqrt{2}}{2}\theta(I_n + I_p). \quad (5.27)$$

Since there are two equations with two unknowns, for a given $I_p (or I_n)$, g_{mT} can be calculated. Let us define the percentage error of g_{mT} caused by the mobility degradation as

$$ER_{mob} = \frac{g_{mT} - G_m}{G_m}; \quad (5.28)$$

this is approximately proportional to $(I_n + I_p)/(\sqrt{I_n} + \sqrt{I_p})$. ER_{mob} was calculated for $I_{nmax} = I_{pmax} = 20\mu A$ and $K = 200\mu A/V^2$ as I_p was varied from 0 to $20\mu A$. The results are shown in Figure 5.16 where three curves corresponding to different values of θ can be found. As expected, a larger θ results in a larger error both in its magnitude and in its variation. Also as expected, the shapes of the curves resembles the letter "U" drawn upside-down; this is because for $\sqrt{I_p} + \sqrt{I_n}$ being more or less constant, the sum of I_n and I_p is minimum when $I_n = I_p$.

Body effect causes the threshold voltage of the transistor to increase with increasing source to bulk voltage. The drain current and the gate to source voltage can still be given by (2.1) and (2.2), except that V_T is no longer a constant and is given by[56]

$$V_T = V_{TO} + \gamma[\sqrt{2\phi_F + V_{SB}} - \sqrt{2\phi_F}] \quad (5.29)$$

where γ is the bulk threshold parameter, ϕ_F is the surface potential at strong inversion, and V_B is the bulk voltage: V_{DD} for p-channel and V_{SS} for n-channel transistors. As mentioned at the development stage of the constant-g_m bias circuit, body effect in the bias circuit is cancelled. This can be clearly seen by observing (5.29), and

$$\sqrt{I_{pmax}K_p/4} + \sqrt{I_{nmax}K_n/4} = \sqrt{2I_nK_n} + \sqrt{2I_pK_p} \quad (5.30)$$

is still valid. The transconductance is given by

$$g_m = 2\sqrt{KI}(1 - \frac{dV_T}{dV_{GS}}). \quad (5.31)$$

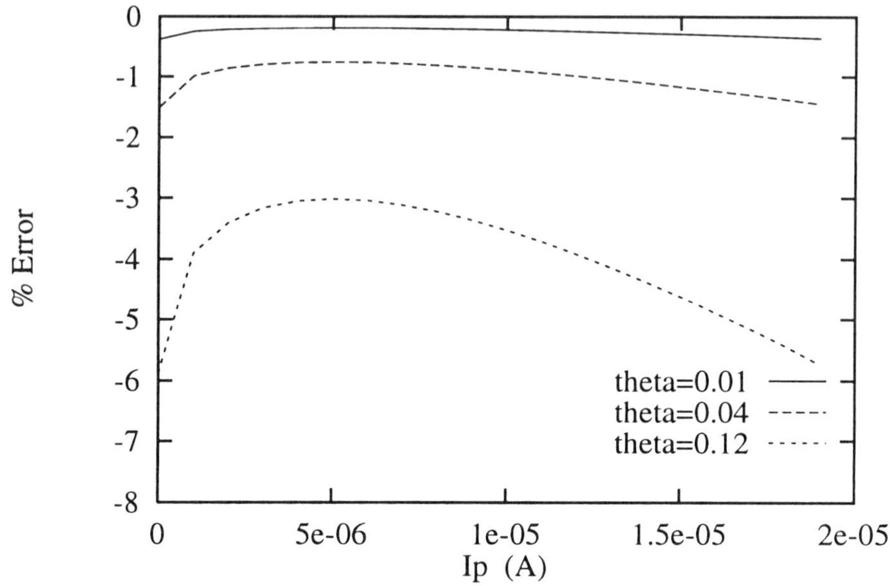

Figure 5.16: The percentage error in g_{mT} caused by the mobility degradation.

In order to evaluate $\frac{dV_T}{dV_{GS}}$, we must be able to express V_S in terms of V_{GS}; however, this requires a specific knowledge of the operation of the transistors. Since we are concerned about g_m of the transistor which is biased with a current through its source terminal, we will assume that the gate voltage is fixed, and then the change in V_{GS} corresponds to the change in $-V_S$, hence g_m is given by

$$g_m = 2\sqrt{KI}(1 + \frac{\gamma}{2\sqrt{2\phi_F + V_{SB}}}). \quad (5.32)$$

Then, g_{mT} of the differential pair is given by

$$g_{mT} = \sqrt{2K_n I_n}(1 + \frac{\gamma_n}{2\sqrt{2\phi_F + V_{SBn}}}) + \sqrt{2K_p I_p}(1 + \frac{\gamma_p}{2\sqrt{2\phi_F + V_{BSp}}}). \quad (5.33)$$

Again, there are two equations (5.32) and (5.33) and two unknowns I_n and I_p. Note that V_{SBn} and V_{BSp} can be calculated using I_n, I_p, and the gate voltage of the differential pair V_{CM}. For the calculations, $I_{pmax} = I_{nmax} =$

$20\mu A$, $K_n = K_p = 200\mu A/V^2$, $2\phi_F = 0.6V$, $V_{TO} = 0.7V$, $V_{DD} = 3V$, $V_{SS} = 0$, and $\gamma_n = \gamma_p = 1$ were used. Figure 5.17 shows the percentage error of g_{mT} as a function of I_p for different values of V_{CM}. The percentage error caused by the body effect is defined as follows.

$$ER_{bod} = \frac{g_{mT} - G_{mr}}{G_{mr}} \tag{5.34}$$

where G_{mr} is the value of g_{mT} for $I_n = I_p = 5\mu A$ and $|V_{SB}| = 0.5V$. The way to interpret the curves is as follows. Since the expression for g_m in (5.32) is derived assuming a constant V_{CM}, any one particular curve by itself does not describe the behavior of g_{mT} as a function of V_{CM}, that is, for a given V_{CM} there is only one possible value for I_p and hence only one value for the error corresponding to the $V_{CM} - I_p$ combination. . We know that I_p decreases with increasing V_{CM}, and for the curve $V_{CM} = 1.0V$, I_p should be large and the error should be about -2%. For $V_{CM} = 2.0V$, I_p is small and the error is -2% again. Once this process is completed for all values of V_{CM}, we will see the error curve shaped as an "up side down U", as was the case for ER_{mob}.

According to the analysis given above, whether the nonideal effect is due to mobility degradation or to the body effect, g_{mT} of the input stage is predicted to be larger at V_{CM} that causes $I_n \approx I_p$ than at V_{CM} that causes one of the currents to be much larger than the other. Combining these results with the analysis performed on the weak inversion operation of the input stage, we realize that g_{mT} can be quite "bumpy". However, calculations of the errors using realistic values of process parameters showed that these errors should be only a small percent.

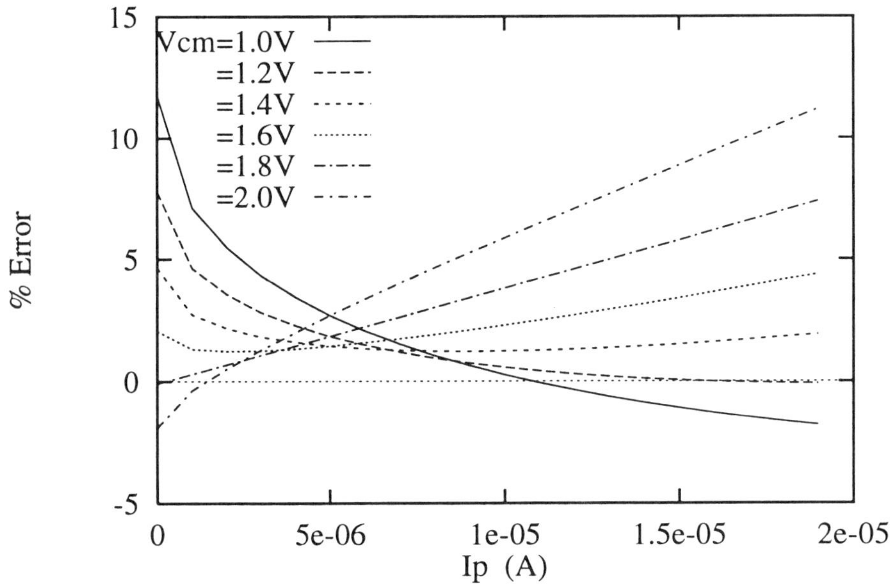

Figure 5.17: The percentage error in g_{mT} caused by the body effect.

Chapter 6

Rail-to-Rail Output Stages

The main goal of this work was to design an input stage which works rail-to-rail and have constant transconductance for the entire common mode range, and it was to be done without the necessity of matching p- and n-channel transistors. Two new circuits which possess these requirements were developed in the previous chapter; however, the effort will be incomplete without the application of these input stages in operational amplifiers. That is, the drain terminals of the differential pairs that were left unconnected in all the circuit diagrams will finally be used as parts of the signal paths in opamps. Thus, the goal of this and the following chapters is to design opamps which enable us to demonstrate the effectiveness of the input stages that are newly developed. This chapter will focus on amplifier architectures and their output stages.

6.1 Design Goals for the Operational Amplifiers

Since we already completed the input stage design, half of our goals are achieved. As there are many ways to complete the opamp design, we will choose the configuration which is widely used. Unless it is necessary to maintain the constant-g_m characteristics of the input stage, no new circuits will be developed, and no novel circuit techniques will be employed to improve the opamp performance such as power supply rejection[57], very high frequency operation, and low noise characteristics. Because we are

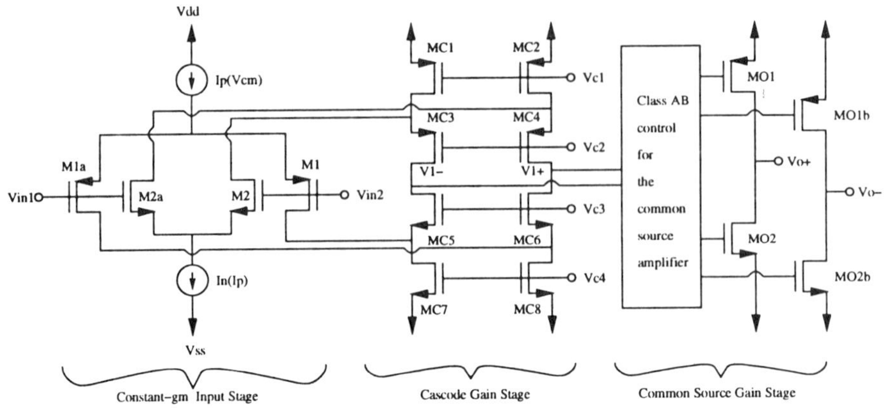

Figure 6.1: Folded-cascode architecture to be used for the opamp with constant-g_m input stage.

going to use a basic architecture, existing improvement techniques of the opamp performances that have already been developed in the literature should easily be applied if necessary.

6.1.1 Operational Amplifier Architecture

The easiest way to obtain a fairly large amount of gain from a single gain stage is to use the folded-cascode architecture shown in Figure 6.1. The constant-g_m input stage is connected to the cascode gain stage consisting of M_{C1} through M_{C8}. V_{1+} and V_{1-} are out of phase, and $V_{1+} - V_{1-}$ corresponds to V_{o1} in the small signal model given in Figure 2.1. For a single stage opamp, the low frequency gain is given by

$$A_{V1} = V_{o1}/V_{in} = g_{mT}R_{o1} \tag{6.1}$$

where R_{o1} is the small signal resistance looking into the drain of M_{C3} or M_{C4}, $V_{in} = V_{in1} - V_{in2}$, and $V_{o1} = V_{1+} - V_{1-}$. R_{o1} is given by the parallel combination of $g_{m4}r_{o2}r_{o4}$ and $g_{m6}r_{o6}r_{o8}$, and then (6.1) becomes

$$A_{V1} = g_{mT}(g_{m4}r_{o2}r_{o4}||g_{m6}r_{o6}r_{o8}). \tag{6.2}$$

Even though a single stage design is capable of providing as much gain as $70dB$ by itself, unless the circuit is only used in switched capacitor applications, any kind of resistive loading will severely reduce the low frequency gain and hence degrade the performance. Moreover, the output nodes of the cascode amplifier need a minimum of two saturation voltages to the rails, and for 3-V operations this makes it difficult to call the cascode amplifier a rail-to-rail output stage. These shortcomings can be solved by adding an output stage consisting of the common source amplifier shown in Figure 6.1. M_{O1} and M_{O2} make up the amplifier and the output node V_{o+} can swing within one saturation voltage to the rails. Also, it acts as a buffer for the cascode output node and hence the external loading of the cascode amplifier is eliminated. In the opamp model of Figure 2.1, the low frequency gain of the second stage is given by $g_{mo}R_{o2}$ where g_{mo} and R_{o2} in Figure 6.1 are $g_{mo1} + g_{mo2}$ and $r_{oo1}||r_{oo2}$, respectively. Then

$$A_{V2} = (g_{mo1} + g_{mo2})(r_{oo1}||r_{oo2}) \quad (6.3)$$

and the total low frequency gain of the opamp is given by $A_{V1}A_{V2}$.

Since both gate terminals of M_{O1} and M_{O2} are driven by signal, they need to be provided with stable bias points. This is achieved by using a class AB control circuit which sets the desired quiescent current for the common source amplifier. The opamp can be made fully differential by simply adding another common source amplifier driven by a signal that is out of phase with respect to the other signal. Note that the same class AB control circuit used for one common source amplifier can also be used for the other. Therefore, there is no need for an extra circuitry or a common mode feedback circuit[58, 59, 60]. The output stage is power efficient since the quiescent current can be set small and the amplifier can provide large current when the output node swing is large. However, when one of the transistors is delivering a lot of current, the other is turned off. In the presence of a high frequency signal with a large amplitude, the time that is required for the transistor to turn back on may introduce distortion in the processed signal. In recently published low voltage opamps(both in CMOS and Bipolar)[42, 43, 37, 38, 51], the class AB circuit is designed in such a way that neither of the output transistors turns off during high current delivering state. The CMOS versions are discussed next and are modified so that they can be used in the opamp architecture given in Figure 6.1.

Table 6.1 shows a summary of opamps to be designed in this work. Opamp 1a and 1b are single-stage single-ended opamps with constant-g_m input stage 1 and 2, respectively. Opamp 1 which has an input stage without the constant-g_m characteristics is also designed so that the performances of the opamps can be compared and the expected improvement in the performance of opamps 1a and 1b can be fully characterized . Opamps 2a and 2b are two-stage design with single-ended output structure and with constant-g_m input stages. They will be compared with opamp 2 which is the same as opamp 2a and 2b except that it has a regular input stage without the constant-g_m characteristics. Opamp 3a and 3b are fully differ-

Table 6.1: Operational amplifiers to be designed in this chapter.

Name	Input Stage	Architecture
Opamp 1	rail-to-rail	single stage, single-ended
Opamp 1a	input stage 1	single stage, single-ended
Opamp 1b	input stage 2	single stage, single-ended
Opamp 2	rail-to-rail	two stage, single-ended
Opamp 2a	input stage 1	two stage, single-ended
Opamp 2b	input stage 2	two stage, single-ended
Opamp 3a	input stage 1	two stage, fully differential
Opamp 3b	input stage 2	two stage, fully differential

ential versions of opamp 2a and 2b, respectively. Chapter VII is devoted to single-ended opamps while Chapter VIII will discuss the two-stage opamps.

6.1.2 Existing CMOS Output Stages With Class AB Control

Class AB amplifiers have two complementary transistors so that at least one of them is operating at all times and the region in which both transistors are operating is fairly wide. A bias circuit for the class AB amplifier must establish a good quiescent condition and moreover it should avoid high frequency performance deterioration by maintaining a certain amount of current for the transistor that would be turned off. The I-V curves of the output stage possessing such characteristics is shown in Figure 6.2 where I_{op} and I_{on} are the currents in the p- and n-channel transistors of the common source amplifier, respectively.

The first CMOS output stage with the characteristics of Figure 6.2 is shown in Figure 6.3[37]. Transistors M_{O1} and M_{O2} form the common source amplifier and they are driven by signals V_{2+} and V_{1+} which are in phase. Note that in a regular folded-cascode architecture as in Figure 6.1, the signals at the drains of M_{C1} and M_{C2} are out of phase and hence so are the signals at the drains of M_{C3} and M_{C4}. Thus, in order to use the output stage in Figure 6.3, the differential input pairs must be modified, but since we would like to stay as close as possible to the conventional design, we will not use the approach taken here. Nevertheless, we will study the operation of this class AB circuit as a starting point toward the design of a similar circuit which is suited to be placed in the architecture shown in Figure 6.1.

When V_{1+} and V_{2+} are increased toward V_{DD}, the gate to source voltage of M_{O2} increases and hence I_{on} increases. At the same time the source to

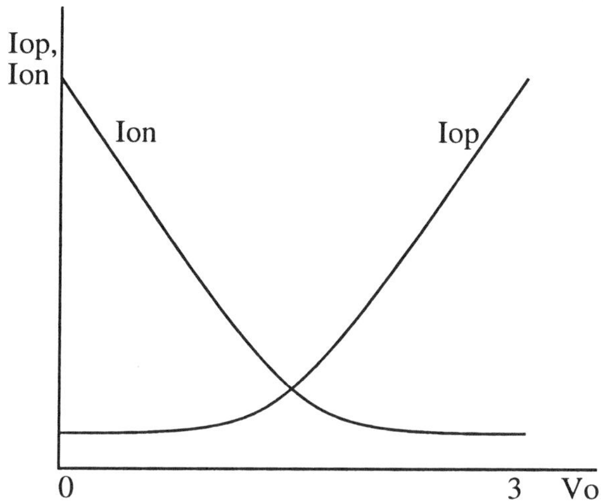

Figure 6.2: Desired I-V characteristics of the output stage with class AB control.

gate voltage of M_{O1} decreases and hence I_{op} decreases. Then the current will flow into the load at the output node and V_o moves toward the negative supply rail. However, if the increase in V_{1+} and V_{2+} required to change the output voltage is too large, the source to gate voltage of M_{O1} will be small enough to turn it off. Similarly, M_{O2} will turn off when V_o is too large. Transistors M_{O5} through M_{O12} prevent the transistors from turning off by monitoring I_{op} and I_{on} and feeding the information back into the signal paths of the opamp.

Transistors M_{O7} and M_{O10} carry a copy of I_{on} and I_{op}, respectively. The gate voltages of the p-channel differential pair consisting of M_{O8} and M_{O9} are given by

$$V_{GO8} = I_{on} R_a + V_{SS} \quad (6.4)$$

$$V_{GO9} = I_{op} R_a + V_{SS}. \quad (6.5)$$

Note that when I_{on} becomes small compared to I_{op}, V_{GO8} will be less than V_{GO9} and a larger portion of I_a will flow into M_{O8}. If I_{on} is further decreased, all of I_a flows into M_{O8} and then M_{O9} turns off, and at this point V_a is only influenced by M_{O8} and hence by I_{on}. That is V_a will be pulled toward V_{SS} by decreasing I_{on}. Actually, since as far as the M_{O8}-M_{O9} pair is concerned, I_{op} and I_{on} are interchangeable and exactly the same behavior will be seen when I_{op} is very small, *i.e.*, V_a is pulled toward V_{SS} by the reduction of I_{op}. So whether I_{op} or I_{on} becomes small, V_a

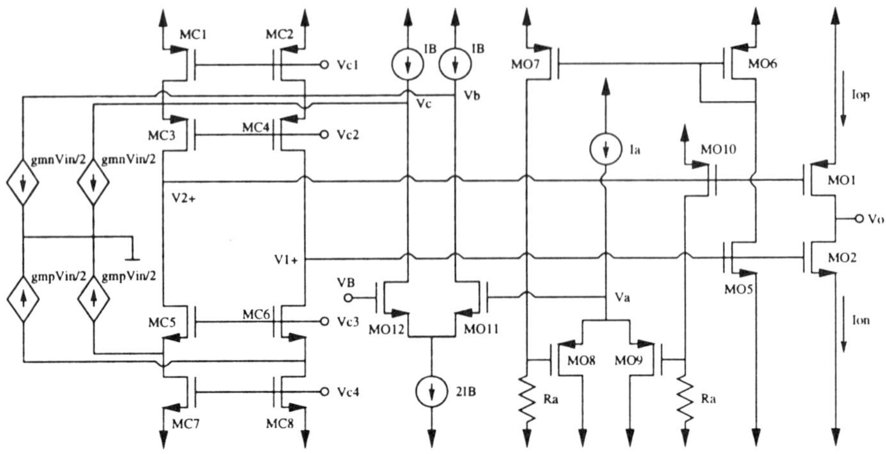

Figure 6.3: Class AB output stage which prevents output transistors from turning off in the presence of a large signal.

will move toward V_{SS}. V_a is one of the input voltages for the n-channel differential pair consisting of M_{O11} and M_{O12}. The other input V_B is a constant and then the operation of this pair is determined by V_a. Let us assume that V_o is to be increased and V_{1+} and V_{2+} are decreased. Then I_{on} becomes small and this will reduce V_a relative to V_B and then V_b increases and V_c decreases. These actions will respectively increase V_{1+} and decrease V_{2+}. The increase in V_{1+} will help M_{O2} stay operating and the decrease in V_{2+} will help M_{O1} deliver more current to the output load. M_{O1} and M_{O2} are affected exactly in the same manner when I_{op} is becoming small.

An alternate implementation of the output stage with the characteristics of Figure 6.2 is shown in Figure 6.4[38]. Note that the cascode stage is again driven by signals that are in phase. The differential pair consisting of M_{O13} and M_{O14} provides the feedback signal to the cascode stage which in turn affects the operation of M_{O1} and M_{O2}. The input signal to this differential pair is $V_d - V_a$ and is given by

$$V_d - V_a = V_{SGO15} - V_{SGO16} + V_{SGO10} - V_{SGO11}. \tag{6.6}$$

The currents flowing in transistors M_{O10}, M_{O11}, M_{O15}, and M_{O16} are

$$I_{O10} = I_{on} - I_{min}/2 \tag{6.7}$$
$$I_{O11} = I_{on} + I_{op} - I_{min} \tag{6.8}$$
$$I_{O15} = I_{op} - I_{min}/2 \tag{6.9}$$
$$I_{O16} = I_{min}/2. \tag{6.10}$$

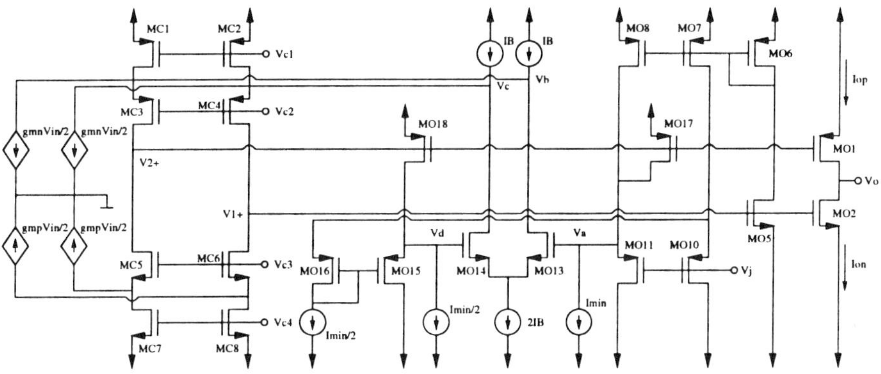

Figure 6.4: Alternate version of class AB output stage which prevents output transistors from turning off in the presence of a large signal.

Assuming they are all in the saturation region, (6.6) is rewritten as follows.

$$\sqrt{K_p}(V_d - V_a) = \sqrt{I_{op} - I_{min}/2} + \sqrt{I_{on} - I_{min}/2} - \sqrt{I_{min}/2} - \sqrt{I_{on} + I_{op} - I_{min}} \tag{6.11}$$

At the quiescent condition, I_{op} and I_{on} are equal, that is, $I_{op} = I_{on} = I_Q$. Then (6.11) can be rewritten as

$$\sqrt{K_p}(V_d - V_a) = (2 - \sqrt{2})\sqrt{I_Q - I_{min}/2} - \sqrt{I_{min}/2}. \tag{6.12}$$

Now let us assume that V_o is swinging towards V_{DD} and I_{op} becomes much larger than I_Q and I_{on} becomes much lower than I_Q, *i.e.*, $I_{op} >> I_Q >> I_{on}$. Then both the first and the last term in the right hand side of (6.11) are dominated by I_{op} and hence they approximately cancel each other and we get

$$\sqrt{K_p}(V_d - V_a) = \sqrt{I_{on} - I_{min}/2} - \sqrt{I_{min}/2}. \tag{6.13}$$

Since I_{on} is much less than I_Q, (6.13) is smaller than (6.12), that is, when I_{on} becomes small, $V_d - V_a$ is also reduced. This change in the input signal for the differential pair $M_{O13} - M_{O14}$ results in V_b moving toward V_{SS} and V_c toward V_{DD}. Then V_{1+} is increased in order to keep M_{O2} operating and V_{2+} is reduced to supply more current to the output load. Since I_{op} and I_{on} are interchangeable in (6.11), the same actions will take place when I_{op} is the current that is becoming small.

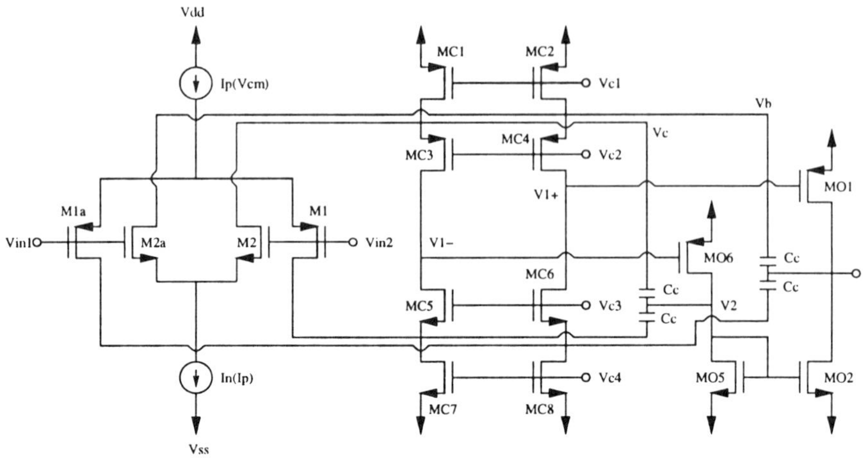

Figure 6.5: Opamp with the modified output stage.

6.2 Modified Class AB Output Stage

As mentioned, we would like to use the traditional opamp architecture shown in Figure 6.1 which has out of phase signals at the output of the cascode stage. However, the output stages that were shown in the previous section were designed based on the availability of the in phase signals that can be directly used to drive the common source amplifier. First, an output stage suitable for the design shown in Figure 6.1 will be studied. Then a class AB control circuit that controls the bias currents of the output transistors is discussed.

6.2.1 The Output Stage

Since the two transistors in the common source amplifier must be driven by signals that are in phase with each other, in order to use the structure given in Figure 6.1, V_{1+} or V_{1-} has to be inverted before it can drive the common source amplifier. Figure 6.5 shows the circuit where V_{1-} is inverted before it reaches the output transistor M_{O2}. Intuitively, one would think that the extra inverting stage consisting of M_{O5} and M_{O6} will introduce an extra pole and degrade the frequency response of the opamp; and this may lead to the conclusion that such circuit should be avoided. However, we will show that with proper frequency compensation, the effect of the extra stage can be minimized and a good response can be obtained.

A small signal circuit model of the opamp in Figure 6.5 is shown in

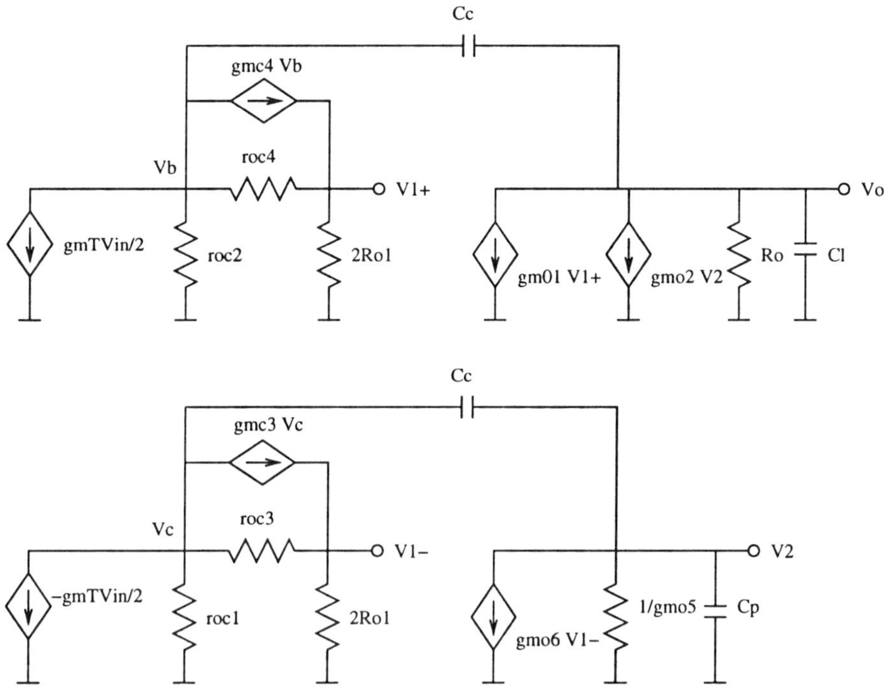

Figure 6.6: Small signal model of the opamp with a modified output stage.

Figure 6.6 where $g_{mXX}$and r_{OXX} are the transconductance and the small signal resistance of a transistor XX, respectively, R_{O1} is the resistance looking into nodes V_{1+} and V_{1-}, C_P is the parasitic capacitance at the gates of M_{O2} and M_{O5}, R_O is the total resistance at the outout node, and C_L is the output capacitance. The transfer function of this opamp model is given by

$$H(s) = \frac{-\frac{g_{mT}}{2}(sC_C - g_{mo1}g_{mc4}r_{oc4})}{s^2C_CC_L + sC_C(g_{mo1}g_{mc4}r_{oc4} + 1/R_O) + \frac{1}{r_{oc2}R_O}}(1+E(s)) \quad (6.14)$$

where

$$E(s) = \frac{g_{mo2}(sC_C + 1/r_{oc2})}{s^2C_CC_P + sC_C g_{mo6}g_{mc3}r_{oc3} + g_{mo5}/r_{oc1}}. \quad (6.15)$$

In the above equations, it was assumed that g_{mo1} and g_{mo6} are approximately the same. Note that if V_{1-} was already in phase with V_{1+} and did not need to be inverted and if it was directly connected to the output node, one side of all four compensation capacitors will be connected to V_O. Then the part of the transfer function(of the opamp with V_{1-} and V_{1+} in phase) that is related to M_{O1} and M_{O2} will be approximately the same. Let $H_O(s)$ be the transfer function of the circuit in which V_{1-} and V_{1+} are already in phase, then the part of (6.14) that is related to M_{O1}(or g_{mo1}) is $H_O(s)/2$, and it can be shown that (6.14) can be written in the form

$$H(s) = \frac{H_O(s)}{2}(1 + E(s)). \tag{6.16}$$

The transfer function of the type given by $H_O(s)$ has good frequency characteristics and $E(s)$ can be looked at as the error term that is caused by the nonideal frequency response of the inverting circuit consisting of $M_{O5} - M_{O6}$. Since $g_{mo2} = g_{mo5}$ and $r_{oc1} = r_{oc2}$, $E(s)$ at low frequencies is unity and becomes zero at high frequencies. That is, $H(s) = H_O(s)$ at low frequencies and it is halved at high frequencies. Let us now study the $1 + E(s)$ term in (6.14) to better understand its effect. We can write

$$1 + E(s) \approx \frac{s^2 C_C C_P + s C_C g_{mo6} g_{mc3} r_{oc3} + 2 g_{mo5}/r_{o1}}{s^2 C_C C_P + s C_C g_{mo6} g_{mc3} r_{oc3} + g_{mo5}/r_{o1}} \tag{6.17}$$

$$\approx \frac{(s + 2/r_{oc1} r_{oc3} g_{mc3} C_C)(s + g_{mo6} g_{mo3} r_{oc3}/C_P)}{(s + 1/r_{oc1} r_{oc3} g_{mc3} C_C)(s + g_{mo6} g_{mo3} r_{oc3}/C_P)}. \tag{6.18}$$

Thus, $1 + E(s)$ has two zeroes and two poles where one of the poles is approximately cancelled by one of the zeroes. These poles and zeroes are at

$$P_{E1} = Z_{E1}/2 = -\frac{1}{r_{oc1} r_{oc3} g_{mc3} C_C} \tag{6.19}$$

$$P_{E2} = Z_{E2} = -\frac{g_{mo6} g_{mc3} r_{oc3}}{C_P} \tag{6.20}$$

and $1 + E(s)$ is expressed as

$$T(s) = \frac{s + 2a}{s + a} \tag{6.21}$$

with its magnitude and the phase given as

$$|T(s)| = \sqrt{\frac{\omega^2 + 4a^2}{\omega^2 + a^2}} \tag{6.22}$$

$$\phi_{T(s)} = tan^{-1}\left[\frac{\omega a}{-2a^2 - \omega^2}\right]. \tag{6.23}$$

The maximum phase deviation in $T(s)$ is -20° and it occurs at $\omega = \sqrt{2}a$; at frequencies that are much smaller or larger than $\sqrt{2}a$, the phase error

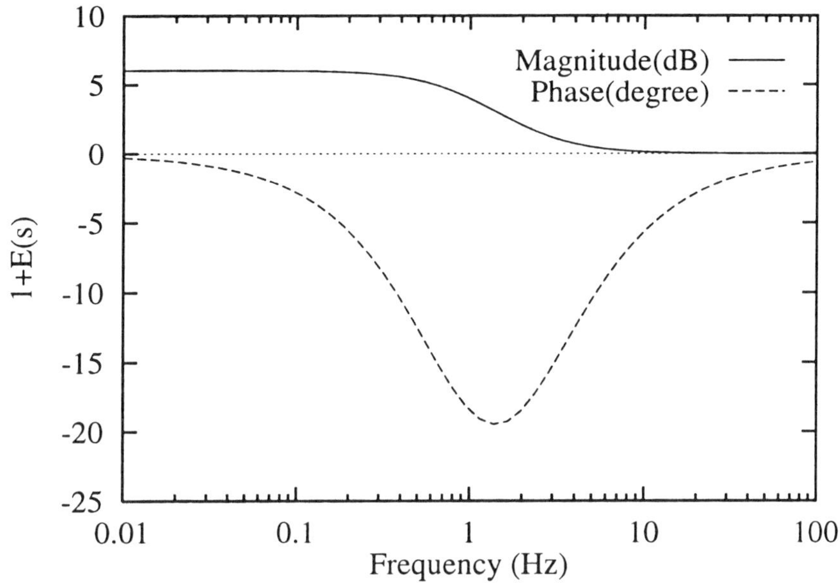

Figure 6.7: The magnitude and the phase response of $\frac{s+2}{s+1}$.

is zero. Figure 6.7 shows $|T(s)|$ and $\phi_{T(s)}$ with the constant a normalized to one. We realize that if P_{E1} is at a frequency much less than the gain-bandwidth product, GBW, and larger than the dominant pole of the opamp, both the magnitude and the phase response of $H(s)$ will be very similar to that of $H_O(s)$. $H_O(s)$ has two poles and a zero and they are at the following frequencies:

$$Z_O = \frac{g_{mo1} g_{mc4} r_{oc4}}{C_C} \tag{6.24}$$

$$P_{O1} = -\frac{1}{C_C R_O r_{oc2} r_{oc4} g_{mo1} g_{mc4}} \tag{6.25}$$

$$P_{O2} = -\frac{g_{mo1} g_{mc4} r_{oc4}}{C_L} \tag{6.26}$$

Note that the low frequency gain of $H_O(s)$ is given by

$$A_{ODC} = g_{mT} g_{mo1} g_{mc4} r_{oc2} r_{oc4} R_O \tag{6.27}$$

and thus GBW of $H_O(s)$ is given by

$$GBW_O = g_{mT}/C_C \tag{6.28}$$

as expected, it is much larger than P_{E1}. The opamp model of Figure 6.6 was simulated and its frequency responses are compared with the responses of $H_O(s)$ which were obtained by setting $g_{mo2} = 0$ and by making g_{mT} twice as large as g_{mT} of $H(s)$. The results are shown in Figure 6.8 and we can clearly see the $6-dB$ gain reduction at higher frequencies and a phase deviation at mid-band in the frequency response of $H(s)$. Actually, because of the $6-dB$ reduction at high frequencies, the unity gain frequency, f_U, of $H(s)$ is half of f_U of $H_O(s)$ for a given set of compensation capacitors. This means that for the same f_U, C_C of the modified output stage needs to be only half the size required for $H_O(s)$; however, this could result in a small reduction in the phase margin, but the amount will depend on the exact locations of poles and zeros. In the case of the circuit used to create Figure 6.8, f_U of $8MHz$ with 87° phase margin is obtained with $C_C = 2pF$ for $H_O(s)$ whereas the same f_U is obtained with $C_c = 1pF$ but the phase margin is reduced to 80° for $H(s)$. Thus, we can actually take advantage of the deviations caused by the insertion of an extra inverting stage between the cascode output node and the common source amplifier. That is, since capacitors occupy a large space, reducing the size of C_c by a factor of two can amount to a significant saving in the chip area. Note that in the opamp model given in Figure 6.6, the contribution of the n-side and the p-side were lumped together. That is, g_{mn} and g_{mp} were lumped into g_{mT}, also, the small signal generated in the n- and p-channel differential pairs were lumped together and injected to the left side and the right side of the cascode stage at nodes V_b and V_c. Thus, the compensation capacitors, C_C, in Figure 6.6 actually represents the lumped sum of the compensation capacitances connected to the n- and p-side of the cascode stage. If all C_C's in Figure 6.5 were $2pF$, then, and because of the reason given above, C_C's in Figure 6.6 would be $4pF$.

6.2.2 The Class AB Control Circuit

Since the class AB control circuits shown in Figure 6.3[37] and Figure 6.4[38] were designed for the case in which the output nodes of the cascode stage are carrying signals that are in of phase with each other, the circuit must be modified before it can be placed in the opamp of Figure 6.5. The design method used here will be based on the circuit of Figure 6.4; however, it will be shown that the design equation given in (6.11) can be simplified and hence the circuit itself will be slightly simpler. The output stage with the

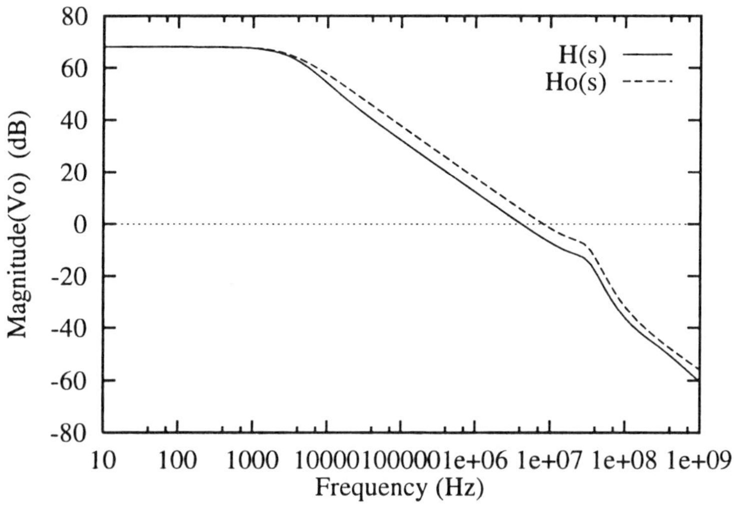

(a)

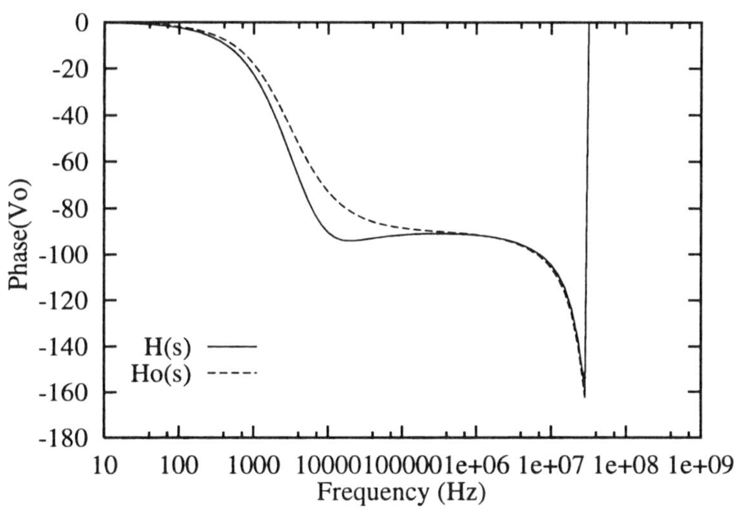

(b)

Figure 6.8: Frequency response of the small signal model of the opamp with modified output stage. (a) Magnitude response. (b) Phase response.

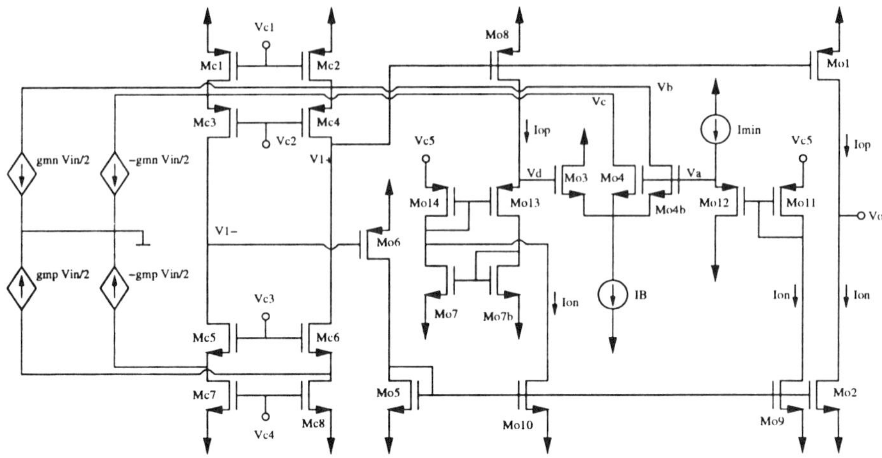

Figure 6.9: Modified class AB output stage.

modified class AB control circuit is shown in Figure 6.9. The differential amplifier consisting of M_{O3}, M_{O4}, M_{O4b} provides the feedback signal to the cascode stage which in turn drives the common source amplifier. Note that V_c and V_b are in phase with each other because M_{O4} and M_{O4b} are both driven by V_a. In the presence of a large signal at the output node, we would like V_{1+} and V_{1-} to move toward V_{SS} in order to prevent M_{O1} or M_{O2} from turning off. Then V_b and V_c should also move toward V_{SS}, and this will happen if $V_d - V_a$ is decreased. Starting from V_d and summing the gate to source voltages of M_{O13}, M_{O14}, M_{O11}, and M_{O12} and going to V_a, we can write

$$V_d - V_a = V_{SGO13} + V_{SGO11} - V_{SGO14} - V_{SGO12}. \tag{6.29}$$

The currents flowing in the transistors M_{O11}, M_{O12}, M_{O13}, and M_{O14} are given by

$$I_{O11} = I_{on} \tag{6.30}$$
$$I_{O12} = I_{min} \tag{6.31}$$
$$I_{O13} = I_{op} \tag{6.32}$$
$$I_{O14} = I_{on} + I_{op} \tag{6.33}$$

then (6.29) can be written as

$$\sqrt{K_p}(V_d - V_a) = \sqrt{I_{op}} + \sqrt{I_{on}} - \sqrt{I_{min}} - \sqrt{I_{op} + I_{on}}. \tag{6.34}$$

At the quiescent condition $I_{op} = I_{on} = I_Q$ and (6.34) becomes

$$\sqrt{K_p}(V_d - V_a) = (2 - \sqrt{2})\sqrt{I_Q} - \sqrt{I_{min}}. \tag{6.35}$$

If the output voltage swings close to V_{DD}, I_{op} becomes very large and then $I_{op} >> I_Q > I_{on}$ is true and the last term in the right hand side of (6.34) is dominated by I_{op} and hence it will approximately cancel the first term. Then (6.34) becomes

$$\sqrt{K_p}(V_d - V_a) = \sqrt{I_{on}} - \sqrt{I_{min}}. \tag{6.36}$$

Thus, when $\sqrt{I_{on}}$ becomes smaller than $(2 - \sqrt{2})\sqrt{I_Q}$, (6.36) becomes smaller than (6.35) and this implies that when I_{on} reaches some minimum level of current, $(V_d - V_a)$ is decreased as desired. The same action will take place when I_{on} becomes large and I_{op} reaches the minimum value. The feedback mechanism tries to maintain the expressions given in (6.34) through (6.36) equal to zero. Then from (6.35) and (6.36), we can find that $I_Q \approx 3I_{min}$ and the minimum value of I_{op} and I_{on} is I_{min}. The modified output stage of Figure 6.9 was simulated with I_{min} of $5\mu A$ in a unity gain buffer configuration. M_{O1} was made twice as large as M_{O8}, and M_{O2} was made twice as large as M_{O9} and M_{O10}. Thus, we expect I_Q to be about $30\mu A$ and that is what we see in Figure 6.10(a). The input-output characteristics of the opamp connected as a buffer is shown in Figure 6.9(b) and it can be seen that the output follows the input from about 0.1V to 2.9V; and this is sufficient to be called a rail-to-rail output stage.

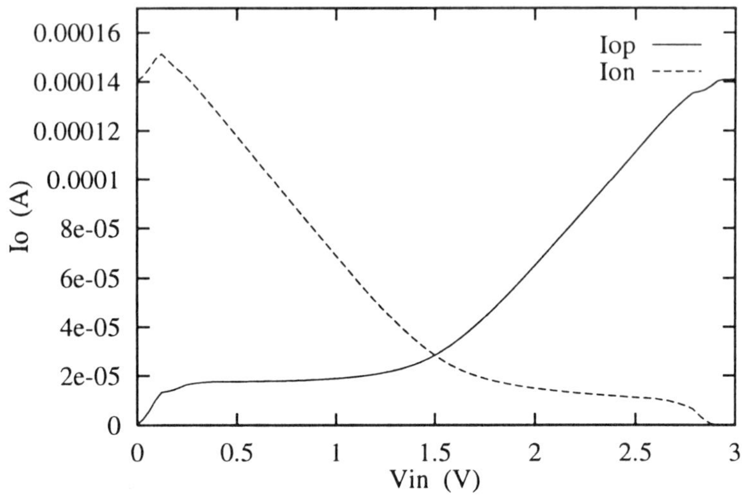

(a)

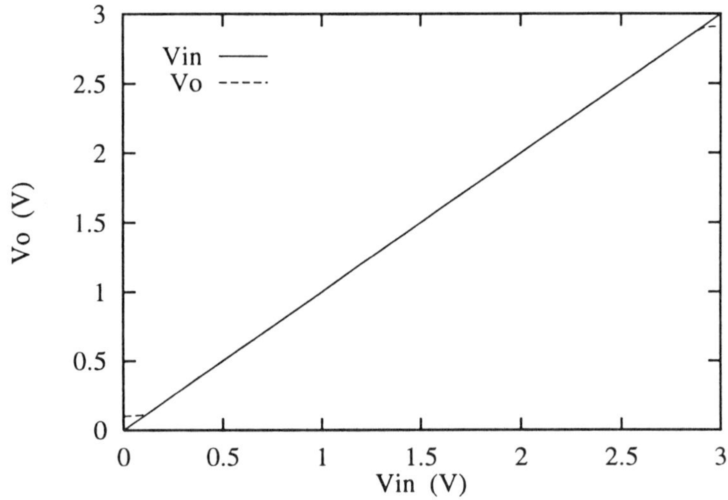

(b)

Figure 6.10: Simulation results of the modified output stage with ideal input stage in a unity buffer configuration. (a) Output currents. (b) Vin-Vo characteristics.

Chapter 7

Single-Stage Operational Amplifiers

In this chapter, we will complete the design of three single-stage opamps which have rail-to-rail input common mode range but without the rail-to-rail output range. Simulation results of the opamps are provided and their performances are discussed.

7.1 Opamp 1: A Simple Folded-Cascode Opamp

Opamp 1 is the simplest opamp designed here, and its performance will be compared with the performance of other opamps. The circuit is shown in Figure 7.1. The opamp output node is the output of the cascode stage and thus the swing in each direction is limited as can be seen in the simulation result shown in Figure 7.2(a). Opamp 1 was connected in a unity gain buffer configuration and the input voltage, V_{in} connected to V_{in1}, was swept from 0 to 3V for $V_{DD} = 3V$ and $V_{SS} = 0V$. The linear output range is from $0.5V$ to $2.3V$. Figure 7.2(b) shows the input stage transconductance of the opamp 1. There are two observations to be made from this plot; first, for input voltages which result in nonlinear transfer characteristics, the virtual short between V_{in1}(positive input) and V_{in2}(negative input) no longer exists and hence the common mode current in M_1 and M_{1a}, or in M_2 and M_{2a}, are not the same anymore. In the plot, g_{mT+} and g_{mT-} are the input stage transconductance of the $V_{in1}-side$ and the $V_{in2}-side$, respectively. For V_{in} near zero, $V_o > V_{in}$ or equivalently, $V_{in2} > V_{in1}$, M_1 receives more current

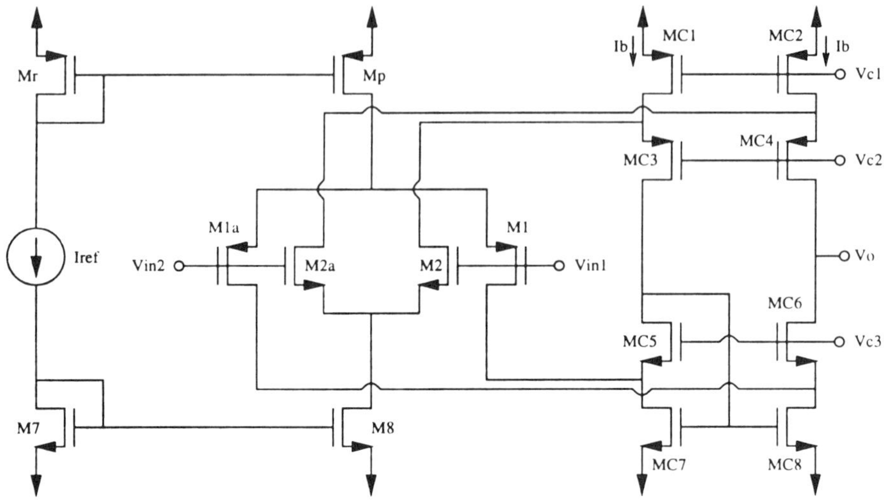

Figure 7.1: Opamp 1: A single stage opamp with rail-to-rail input range.

from M_p than M_{1a}. Because of this and the fact that the p-channel pair is operating at this point, the transconductance seen from the $V_{in1} - side$ is larger. For V_{in} near V_{DD}, $V_{in1} > V_{in2}$, and hence M_2 is conducting more current than M_{2a} and this again results in larger transconductance from the $V_{in1} - side$. The second point to be made is that because there is no circuit to maintain g_{mT} constant we see a significant change in g_{mT} as V_{in} is varied from rail to rail. Figure 7.3 shows the open loop small signal frequency response of the opamp 1 with different common mode input voltages and we can clearly see the dependence of both the unity gain frequency and the low frequency differential gain of the opamp on V_{CM}. More detailed simulation results are provided in Table 7.1 which shows the low frequency gain, A_{DC}, the unity gain frequency, f_U, and the phase margin, ϕ_M for V_{CM} swept from 0.5V to 2.5V. For $V_{CM} < .7V$ and $V_{CM} > 2.2V$, A_{DC} drops drastically due to the cascode stage transistors being pushed into the triode region. In the region $0.7V \leq V_{CM} \leq 2.2V$, f_U changes between $1.59MHz$ and $2.7MHz$ for $5pF$ load and between $0.275MHz$ and $0.474MHz$ for $30pF$ load. Next, we will investigate other opamp performance criteria.

CMRR: Figure 7.4 and Table 7.2 show the common mode rejection ratio, $CMRR$, of opamp 1. Simulations were performed with ac signals, v_{cm1}, at the positive input terminal and, v_{cm2}, connected between the negative input and the output terminals. The common mode dc voltage V_{CM} was placed

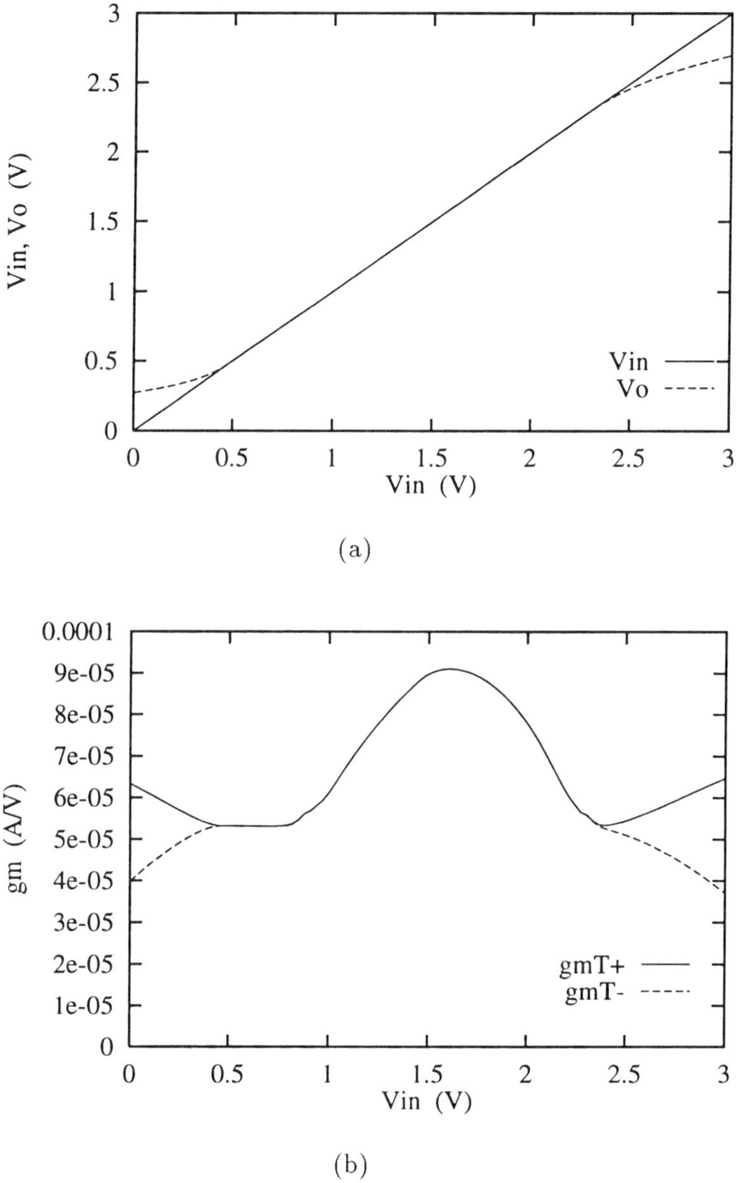

Figure 7.2: Simulation results of opamp 1 in a unity gain configuration: (a) Vin-Vo characteristics, (b) Input stage transconductance.

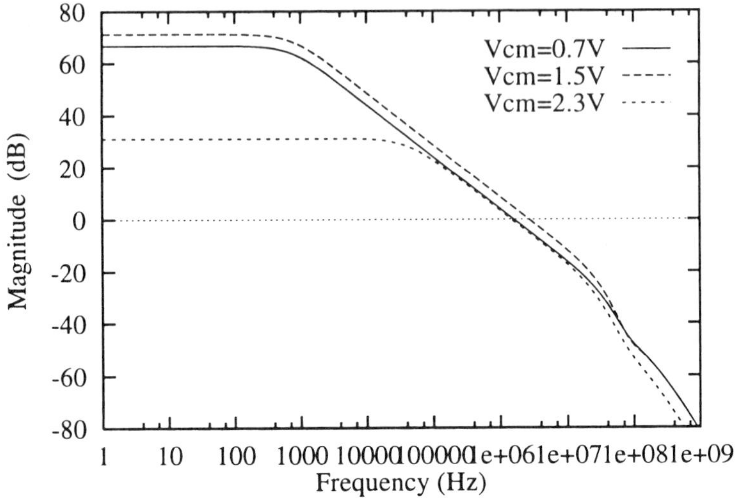

(a)

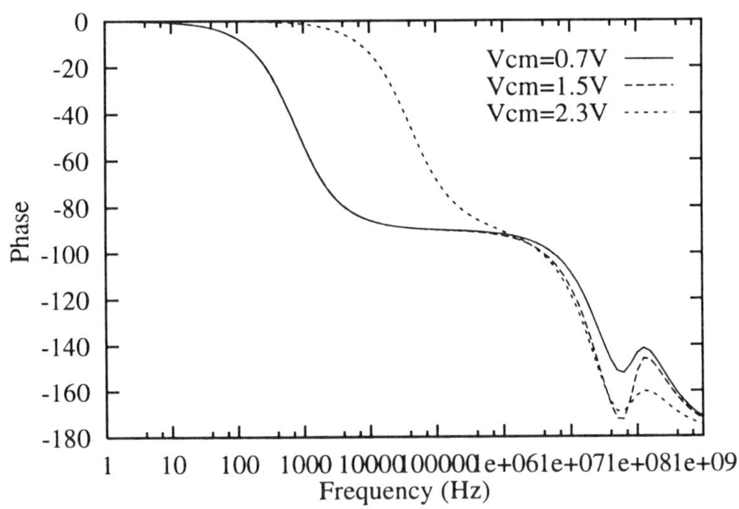

(b)

Figure 7.3: Open loop frequency response simulation of opamp 1: (a) Magnitude response, (b) Phase response.

in series with v_{cm1}. It can then be shown that for $v_{cm1} = v_{cm2} = v_{in}$[61],

$$v_o/v_{in} = 1/CMRR \tag{7.1}$$

as long as the differential gain is much larger than one. Analyzing the small signal circuit model to determine the common mode gain will show that at low frequencies, it is proportional to $1/R_s$ whereas at high frequencies, it is proportional to g_m. R_s is the small signal resistance of the current source which provides I_n or I_p to the differential pairs and g_m is the transconductance of the input transistors. Thus, for $V_{CM} = 1.0$ or $2.0V$, one of the current sources is pushed into the triode region and R_s is significantly reduced. This is the reason for the larger $|CMRR|$ for $V_{CM} = 1.5V$ compared to the value when $V_{CM} = 1.0$ or $2.0V$ at low frequencies. However, note that the range of values of $CMRR$ for different V_{CM} values becomes

Table 7.1: Simulated frequency response of opamp 1.

		$C_L = 5pF$		$C_L = 30pF$	
$V_{CM}(V)$	$A_{DC}(dB)$	$f_U(MHz)$	$\phi_M(^O)$	$f_U(MHz)$	$\phi_M(^O)$
0.5	40	1.40	90	0.275	90
0.6	55	1.59	90	0.275	90
0.7	67	1.68	90	0.275	90
0.8	67	1.59	90	0.275	90
0.9	67	1.68	89	0.301	89
1.0	68	1.83	89	0.327	89
1.1	69	2.04	89	0.356	89
1.2	69	2.25	89	0.388	89
1.3	70	2.40	89	0.422	89
1.4	71	2.57	89	0.460	89
1.5	71	2.67	89	0.460	89
1.6	71	2.70	89	0.474	89
1.7	71	2.67	89	0.474	89
1.8	71	2.62	89	0.474	89
1.9	71	2.52	88	0.474	88
2.0	70	2.35	87	0.400	87
2.1	69	2.10	89	0.360	89
2.2	67	1.83	89	0.327	89
2.3	31	1.48	91	0.270	91
2.4	13	1.10	102	0.255	102
2.5	5	0.90	121	0.214	121

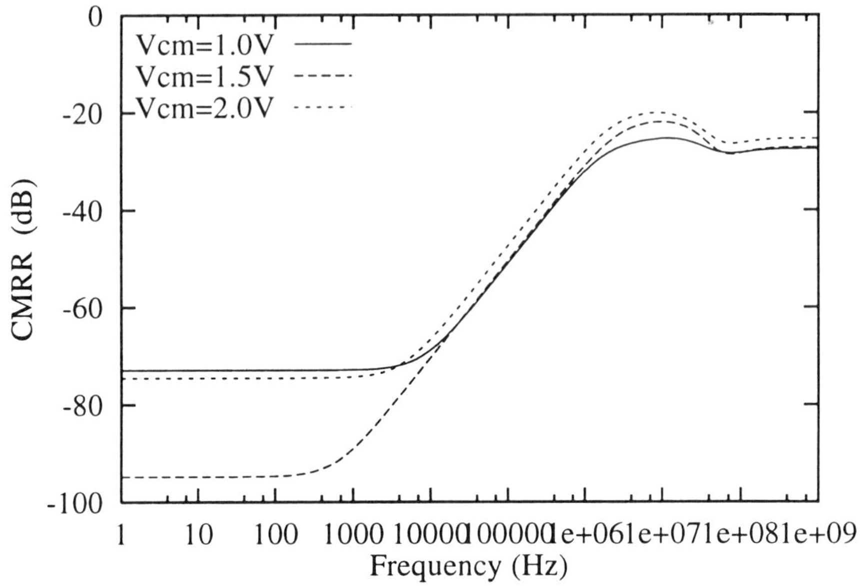

Figure 7.4: $CMRR$ simulation of opamp 1.

smaller at high frequencies. This is because the differences in g_{mT} are less than a factor of two($< 6dB$).

PSRR: The simulation results of the power supply rejection ratio, $PSRR$, of opamp 1 are shown in Figure 7.5 and Table 7.3. $PSRR$ was obtained by applying a small signal ac input at either supply rails of a buffer connected opamp. For both the positive and negative $PSRR$, the largest contribution to the output voltage from the power supplies come from the cascode stage. First, let us consider the negative $PSRR$. The power supply noise that enters the source of M_{C8} reaches the output node with the gain of

$$v_o/v_{sC8} = g_{mC8}R_{o1+} \tag{7.2}$$

where R_{o1+} is the resistance at the output node. The signal that enters through the source of M_{C7} will first travel to the drain terminal of M_{C5} with the gain of $g_{mC7}R_{o1-}/(1 + g_{mC7}R_{o1-})$ where R_{o1-} is the resistance at the drain terminal of M_{C5}. It will then reach the output terminal with

Table 7.2: Simulated common mode rejection ratio of opamp 1.

		$1/CMRR(dB)$	$@(Hz)$				
$V_{CM}(V)$	$C_L(pF)$	DC	$1K$	$10K$	$100K$	$1M$	$10M$
1.0	5	73	73	69	51	32	25
1.5	5	95	88	70	51	31	22
2.0	5	75	74	67	47	28	20
1.0	30	73	73	69	51	41	40
1.5	30	95	88	70	51	38	37
2.0	30	75	74	67	47	36	35

the same gain as in (7.2) but with a sign change. That is

$$v_o/v_{sC7} = -g_{mC8}R_{o1+}\frac{g_{mC7}R_{o1-}}{1+g_{mC7}R_{o1-}} \tag{7.3}$$

and the total gain from V_{SS} through the cascode stage is given by the sum of (7.2) and (7.3). Note that the matching between M_{C7} and M_{C8} is not very important as far as the $PSRR$ is concerned. For the $PSRR$ from V_{DD}, the matching between M_{C1} and M_{C2} is very important since the gain from the source terminal of M_{C1} is proportional to g_{mC1} and the gain from the source terminal of M_{C2} is proportional to g_{mC2}. The simulation results provided here assume no mismatch; however, it was verified that a slight mismatch between M_{C1} and M_{C2} results in a significant degradation in $PSRR$ from V_{DD} whereas the mismatch between M_{C7} and M_{C8} had almost no effect.

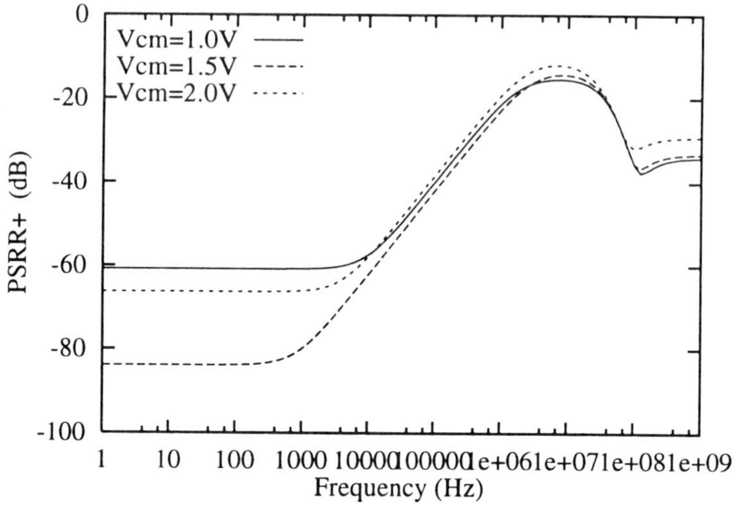

(a)

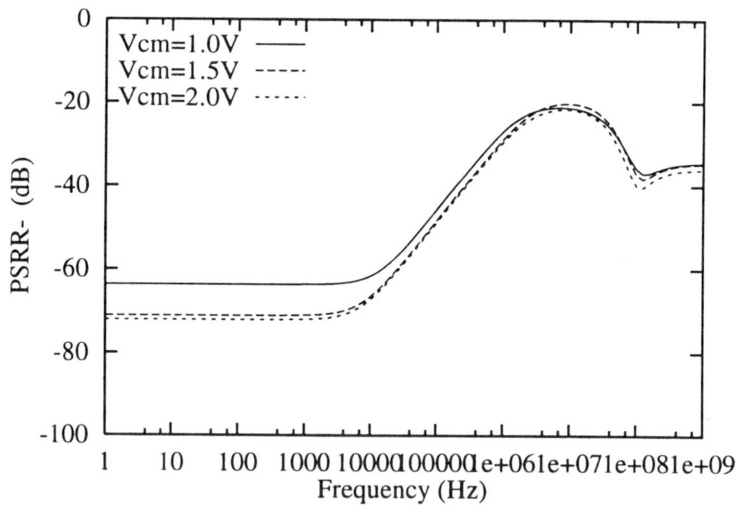

(b)

Figure 7.5: Power supply rejection ratio simulation of opamp 1: (a) Positive supply, (b) Negative supply.

Table 7.3: Simulated power supply rejection ratio of opamp 1.

		$1/PSRR(dB)$ from V_{DD} @(Hz)					
$V_{CM}(V)$	$C_L(pF)$	DC	$1K$	$10K$	$100K$	$1M$	$10M$
1.0	5	60	60	60	47	24	15
1.5	5	84	84	72	49	26	14
2.0	5	66	66	64	45	22	12
1.0	30	60	60	60	47	31	30
1.5	30	84	84	72	49	30	29
2.0	30	66	66	64	45	28	27
		$1/PSRR(dB)$ from V_{SS} @(Hz)					
$V_{CM}(V)$	$C_L(pF)$	DC	$1K$	$10K$	$100K$	$1M$	$10M$
1.0	5	63	63	63	52	30	21
1.5	5	71	71	70	55	32	20
2.0	5	72	72	71	55	33	22
1.0	30	63	63	63	52	37	36
1.5	30	71	71	70	55	37	35
2.0	30	72	72	71	55	38	37

The Settling Time: Table 7.4 shows the 2% settling time of opamp 1 for various input signals. The input $V_{in\ INIT}$ is the initial value of the step input applied at the positive input terminal of the unity gain connected opamp. The input voltage steps up by $V_{in\ STEP}$ and after some time it is stepped down to the initial value. The step size of $0.2V$ was applied for three different values of $V_{in\ INIT}$. Since the small signal settling time, T_S, is approximately proportional to the unity gain frequency, it is expected that T_S for V_{in} close to $1.5V$(where f_U is larger) is less than that for other V_{in} values. Further, note that T_S depends on the steady state final value of V_{in}, that is, for a step input starting at $0.9V$, stepping up to $1.1V$, and stepping back down to $0.9V$, $T_{S\uparrow}$ will depend on the f_U for $V_{CM} = 1.1V$, and $T_{S\downarrow}$ will depend on f_U for $V_{CM} = 0.9V$. This is exactly what can be seen for $0.2V$ steps in Table 7.4. For larger step size, the settling behavior starts with the large signal slewing and then it is followed by the small signal settling. Thus, T_S will gradually increase for larger steps.

Distortion: Table 7.5 shows the distortion analysis of the opamp 1 in the unity gain configuration. It was shown in Chapter II that the variation in the input transconductance, g_{mT}, introduces a distortion and hence opamp 1 should have THD which is dependent on V_{CM}. We can see from the table,

Table 7.4: Simulation results of 2% settling time of opamp 1 with $C_L = 5pF$.

$V_{in\ STEP}(V)$	$V_{in\ INIT}(V)$	$T_{S\uparrow}(\mu sec)$	$T_{S\downarrow}(\mu sec)$
0.2	0.9	0.32	0.38
0.2	1.4	0.23	0.23
0.2	1.9	0.31	0.25
0.4	1.3	0.24	0.23
0.6	1.2	0.24	0.24
0.8	1.1	0.25	0.24
1.0	1.0	0.26	0.34
1.2	0.9	0.29	0.37
1.4	0.8	0.33	0.42
1.6	0.7	0.36	0.43

that THD for $V_{CM} = 1.5V$ and $v_{in} = 0.2V$ is about 4 to 5 times smaller than the case when $V_{CM} = 1.0V$ or $2.0V$, as expected. This observation is valid whether the fundamental frequency of the signal is $10KHz$, $100KHz$, or $200KHz$. However, for larger frequencies, the opamp gain is reduced and the distortion becomes larger.

7.2 Opamp 1a: A Folded-Cascode Opamp With Input Stage 1

Opamp 1a is shown in Figure 7.6; the only difference between opamp 1 and this one is the bias circuit which keeps the input stage transconductance constant. The bias circuit used is shown in Figure 5.9. Figure 7.7(a) shows the dc transfer curve which is identical to that for opamp 1, and Figure 7.7(b) shows g_{mT} of opamp 1a which is significantly more constant than that of opamp 1 within the linear region. The same reasons that are used to explain the difference between the transconductance seen from the $V_{in1} - side$ and the $V_{in2} - side$ at common mode voltages close to the rails for opamp 1 apply to opamp 1a as well.

The simulation results of the small signal frequency response of opamp 1a are shown in Figure 7.8 and in Table 7.6. Figure 7.8(a) shows that the unity gain frequency of opamp 1a is almost constant even when the low frequency gain is severely reduced due to the cascode stage being pushed into the triode region. From Table 7.6, we can see that f_U is fairly constant for $V_{CM} = 0.5V$ to $2.3V$.

Table 7.5: Simulated total harmonic distortion of opamp 1 with $C_L = 5pF$.

$f_O(Hz)$	$V_{CM}(V)$	$v_{in}(V)$	$THD(\%)$
10K	1.0	0.2	0.047
10K	1.5	0.2	0.009
10K	2.0	0.2	0.041
10K	1.5	0.5	0.037
10K	1.5	0.6	0.050
10K	1.5	0.7	0.063
10K	1.5	0.8	0.080
10K	1.5	0.9	0.34
10K	1.5	1.0	1.18
100K	1.0	0.2	0.47
100K	1.5	0.2	0.086
100K	2.0	0.2	0.40
100K	1.5	0.5	0.37
200K	1.0	0.2	0.91
200K	1.5	0.2	0.17
200K	2.0	0.2	0.79
200K	1.5	0.5	0.71

The common mode rejection ratio of opamp 1a is studied in Table 7.7. The rejection ratio should be the worst when V_{CM} is large enough to put M_p into the triode region. Note that M_8 will not go into the triode region even for a small V_{CM} because its gate voltage will be reduced at the same time due to the reduction of the current in M_7. Power supply rejection ratios of opamp 1a are shown in Table 7.8. As mentioned earlier, $PSRR$ of the single stage structure used here depends on the cascode stage and thus, the results given in Table 7.8 are similar to those given in Table 7.3.

The 2% settling time of the opamp 1a is shown in Table 7.9. Since the unity gain frequency of the opamp 1 is larger than that of opamp 1a at $V_{CM} = 1.5V$, T_S for a $0.2V$ step around $1.5V$ is slightly larger for opamp 1a. However, because of the constant-g_m input stage, the range of values for T_S of opamp 1a around different common mode input voltages is smaller than that for opamp 1. For larger steps, T_S increases as expected due to the slew rate limitations; however, we notice that $T_{S\downarrow}$ is significantly less than $T_{S\uparrow}$. The reason for this is explained as follows. The output current, I_O, leaving the output node is given by

$$I_O = I_{2a} - I_2 + I_1 - I_{1a}. \tag{7.4}$$

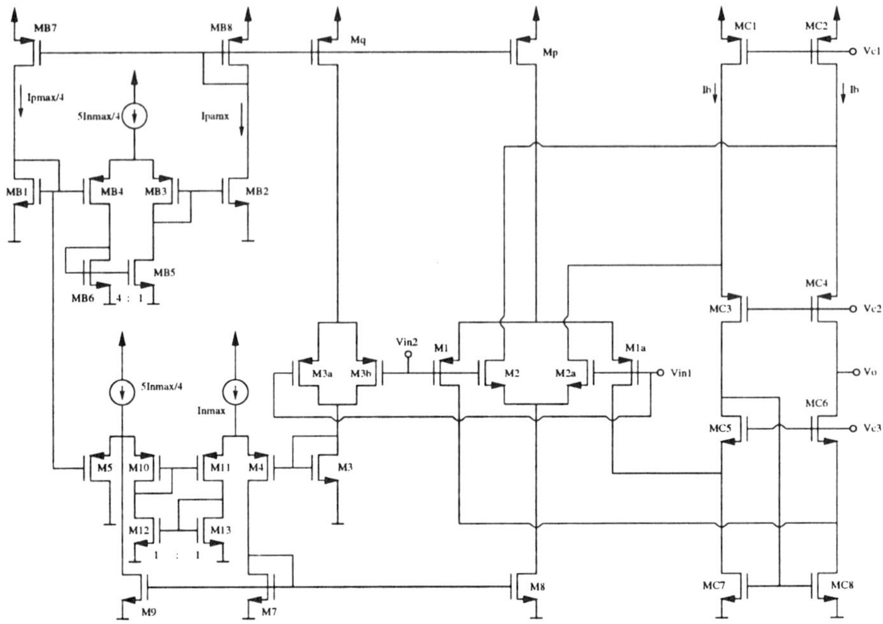

Figure 7.6: Opamp 1a: A single stage opamp with rail-to-rail constant-g_m input stage 1.

At steady state, when V_{in1} is near V_{SS}, V_O and hence V_{in2} will be equal to V_{in1}. Then I_2 and I_{2a} are zero and M_1 and M_{1a} receive their maximum amount. When V_{in1} is stepped toward V_{DD}, the following will take place: I_1 gets larger than I_{1a}, M_p will be pushed into the triode region. Consequently, I_1 will receive all the current M_p provides, but the amount will not be as large as its maximum value.

Since I_2 and I_{2a} were zero to begin with and because of the delay between the time the current in M_p was reduced and the time the constant-g_m bias circuit provides current to M_8, I_2 and I_{2a} are still zero at the first moment of the slewing process. Thus, the current available to charge up the load capacitor for an increasing step is given by

$$I_{O\uparrow} = I_1. \tag{7.5}$$

Now consider the case when V_{in1} and V_{in2} are close to V_{DD} at steady state and a negative step is applied at V_{in1}. In this case, I_1 and I_{1a} are zero and I_2 and I_{2a} are at their maximum value initially. When V_{in1} is stepped down toward V_{SS}, M_2 will receive all the current provided by M_8 and similarly to the other case, M_8 may be pushed into the triode region, so that I_2 will

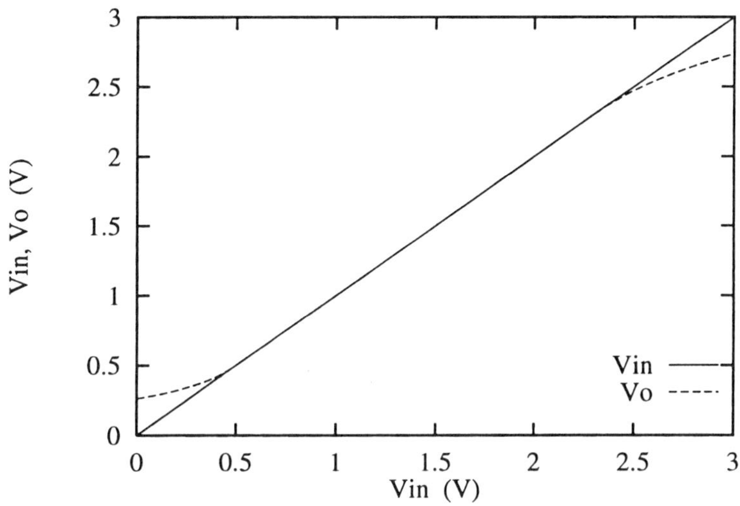

(a)

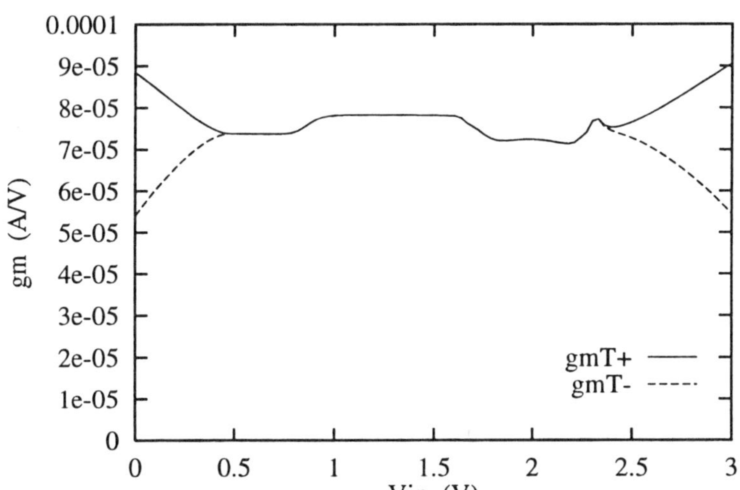

(b)

Figure 7.7: Simulation results of opamp 1a in a unity gain configuration: (a) Vin-Vo characteristics, (b) Input stage transconductance.

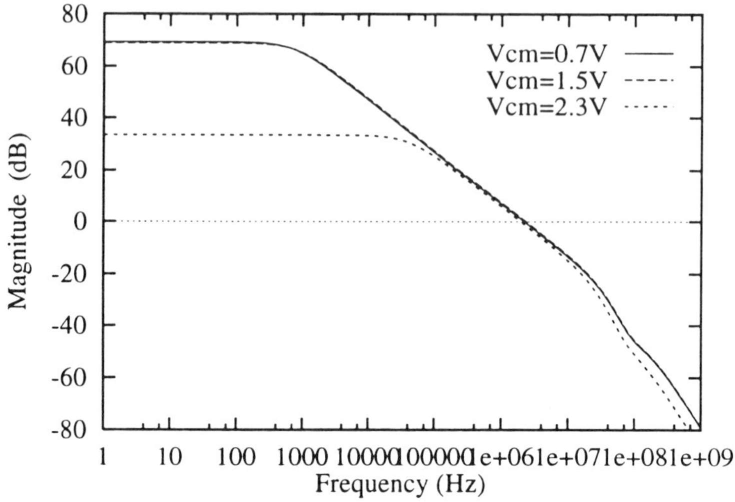

(a)

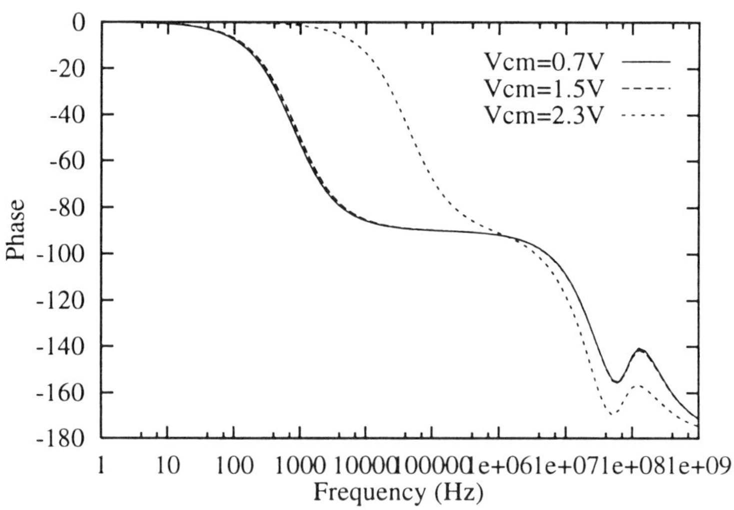

(b)

Figure 7.8: Open loop frequency response simulation of opamp 1a: (a) Magnitude response, (b) Phase response.

Table 7.6: Simulated frequency response of opamp 1a.

		$C_L = 5pF$		$C_L = 30pF$	
$V_{CM}(V)$	$A_{DC}(dB)$	$f_U(MHz)$	$\phi_M(^O)$	$f_U(MHz)$	$\phi_M(^O)$
0.5	41	1.91	87	0.375	90
0.6	56	2.19	86	0.386	89
0.7	69	2.21	86	0.386	89
0.8	69	2.24	86	0.390	89
0.9	70	2.31	86	0.404	89
1.0	70	2.36	86	0.410	89
1.1	70	2.37	86	0.410	89
1.2	69	2.37	86	0.410	89
1.3	69	2.37	86	0.410	89
1.4	69	2.37	86	0.410	89
1.5	69	2.36	86	0.410	89
1.6	68	2.35	86	0.408	89
1.7	68	2.25	85	0.392	89
1.8	68	2.15	84	0.379	89
1.9	68	2.15	83	0.379	89
2.0	68	2.13	82	0.379	89
2.1	69	2.08	82	0.375	88
2.2	69	2.04	82	0.371	88
2.3	33	1.96	85	0.385	89
2.4	16	1.56	95	0.356	98
2.5	8	1.38	110	0.323	112

not be at its maximum value. However, since the gate terminal of M_p is fixed at a certain voltage, as soon as V_{in1} goes toward V_{SS}, M_{1a} will start conducting some current. Then the current available to discharge the load capacitor is given by

$$I_{O\downarrow} = -I_2 - I_{1a}. \tag{7.6}$$

Even though I_2 is less than its maximum value as I_1 in (7.5), due to I_{1a}, the slew rate for the negative step input is less than the slew rate for the positive step input.

The simulated results of the harmonic distortions of opamp 1a are shown in Table 7.10. We see that the range of values of the THD of opamp 1a with a sinusoid of $0.2V$ magnitude is small for different values of the input dc component. This is due to the fact that the input stage has a constant transconductance and thus the results demonstrate the effectiveness of the

Table 7.7: Simulated common mode rejection ratio of opamp 1a.

		$1/CMRR(dB)$ $@(Hz)$					
$V_{CM}(V)$	$C_L(pF)$	DC	$1K$	$10K$	$100K$	$1M$	$10M$
1.0	5	98	90	71	51	31	22
1.5	5	118	90	70	50	31	22
2.0	5	96	80	60	40	21	17
1.0	30	98	90	71	51	39	37
1.5	30	118	90	70	51	39	37
2.0	30	96	80	60	40	29	32

Table 7.8: Simulated power supply rejection ratio of opamp 1a.

		$1/PSRR(dB)$ from V_{DD} $@(Hz)$					
$V_{CM}(V)$	$C_L(pF)$	DC	$1K$	$10K$	$100K$	$1M$	$10M$
1.0	5	63	63	59	42	22	14
1.5	5	80	76	62	42	22	14
2.0	5	63	63	55	36	17	13
1.0	30	63	63	59	42	30	29
1.5	30	80	76	62	42	30	29
2.0	30	63	63	55	36	25	27
		$1/PSRR(dB)$ from V_{SS} $@(Hz)$					
$V_{CM}(V)$	$C_L(pF)$	DC	$1K$	$10K$	$100K$	$1M$	$10M$
1.0	5	69	69	66	49	31	25
1.5	5	69	69	66	49	31	25
2.0	5	69	69	66	49	29	23
1.0	30	69	69	66	49	38	40
1.5	30	69	69	66	49	39	40
2.0	30	69	69	66	49	37	37

Table 7.9: Simulation results of 2% settling time of opamp 1a with $C_L = 5pF$.

$V_{in\ STEP}(V)$	$V_{in\ INIT}(V)$	$T_{S\uparrow}(\mu sec)$	$T_{S\downarrow}(\mu sec)$
0.2	0.9	0.28	0.27
0.2	1.4	0.27	0.26
0.2	1.9	0.35	0.31
0.4	1.3	0.30	0.28
0.6	1.2	0.32	0.27
0.8	1.1	0.35	0.30
1.0	1.0	0.38	0.30
1.2	0.9	0.43	0.32
1.4	0.8	0.48	0.33
1.6	0.7	0.53	0.33

circuit.

7.3 Opamp 1b: A Folded-Cascode Opamp With Input Stage 2

Opamp 1b is shown in Figure 7.9 and its dc transfer characteristics are shown in Figure 7.10. It uses the input stage shown in Figure 5.11. The input-output relation is exactly the same as the one for opamp 1a, whereas the g_{mT} is similar, but with different bumps. Nonetheless, within the region where the input-output characteristic is linear, in both opamp 1a and opamp 1b we have succeeded to obtain g_{mT} much more constant compared to that of opamp 1.

Table 7.11 shows the small signal frequency response of opamp 1b. As it was the case for opamp 1a, the constant-g_m input stage keeps the unity gain frequency fairly constant.

Because the gate voltage terminals of M_p and M_8 are both dependent on the common mode input voltage, these transistors will not go into the triode region and thus their small signal resistance should remain large regardless of V_{CM}. The common mode signal entering the gates of M_{3a} and M_{3b} become a factor as far as the change in $CMRR$, as a function of V_{CM}, is concerned. For $V_{CM} = 1.0V$, M_{3a} and M_{3b} are turned off and hence the rejection should be good as seen in Table 7.12. For $V_{CM} = 1.5V$, M_{3a} and M_{3b} will be on and the common mode signal will propagate through M_r, M_p, to the p-channel differential pair, and also M_r, M_q, eventually to

Table 7.10: Simulated total harmonic distortion of opamp 1a with $C_L = 5pF$.

$fo(Hz)$	$V_{CM}(V)$	$v_{in}(V)$	$THD(\%)$
10K	1.0	0.2	0.007
10K	1.5	0.2	0.011
10K	2.0	0.2	0.011
10K	1.5	0.5	0.037
10K	1.5	0.6	0.041
10K	1.5	0.7	0.045
10K	1.5	0.8	0.052
10K	1.5	0.9	0.18
10K	1.5	1.0	0.86
100K	1.0	0.2	0.063
100K	1.5	0.2	0.11
100K	2.0	0.2	0.11
100K	1.5	0.5	0.37
200K	1.0	0.2	0.12
200K	1.5	0.2	0.21
200K	2.0	0.2	0.23
200K	1.5	0.5	0.73

the n-channel differential pair. Since both differential pairs are operating for this V_{CM}, $CMRR$ is expected to be the worst. For $V_{CM} = 2.0V$, the p-channel pair will be turned off and there will be no common mode signal propagation from the p-channel differential pair side. $PSRR$ is again determined by the cascode stage and the results shown in Table 7.13 are similar to those of opamp 1 and opamp 1a.

The simulation results of the step response of opamp 1b are shown in Table 7.14. The results are similar to those for opamp 1a, except for the increase in $T_{S\downarrow}$ for large input steps. This is because of the dependence of the gate voltage of M_p on V_{CM}. That is, when V_{in1} is suddenly decreased toward V_{SS}, I_{1a} cannot increase until M_b, M_r, and M_p sense the change. This delay will cause I_{1a} to be less significant at the first phase of the discharging process of the output load capacitor.

Table 7.15 shows the results of the harmonic distortion simulation of opamp 1b. The results are similar to those of opamp 1a except that THD is slightly larger when the input sinusoid swings around $1.5V$ than when it swings $1.0V$ or $2.0V$. This is because the biggest bump in g_{mT} occurs when the common mode input voltage is close to $1.5V$.

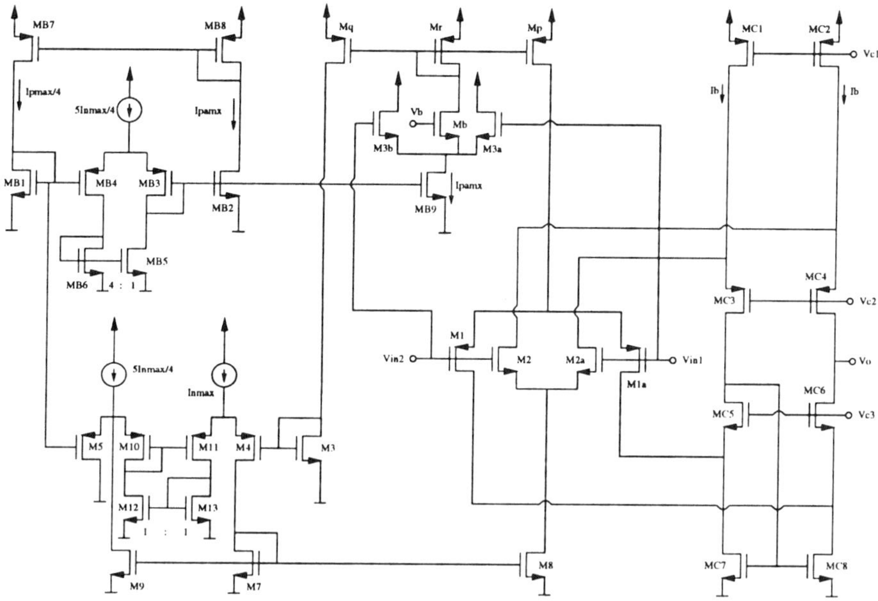

Figure 7.9: Opamp 1b: A single stage opamp with rail-to-rail constant-g_m input stage 2.

Simulated quiescent power dissipation of the opamps are $0.5mW$ and $0.8mW$ for opamp 1 and opamps 1a and 1b, respectively. To compare the performance of the single stage opamps discussed here, the figures given in Tables 7.1, 7.6, and 7.11 are used to compare how constant f_U is as a function of the input common mode voltage. For $V_{CM} = 0.7V$ to $2.2V$, where the cascode stage is operating properly, the maximum and minimum value of f_U, f_{Umax} and f_{Umin}, respectively, are read and their differences are divided by the average value of f_U, f_{Uave} and the percentage deviation in f_U are determined. The results are shown in Table 7.16 which indicates about 50% deviation for opamp 1 and 10 to 15% deviation for the opamps with the constant-g_m input stage. Note that one should exercise care in using this method for determining the deviation in f_U. For example, if $f_{Umax} = 1MHz$ and $f_{Umin} = 0.5MHz$, the maximum value is 100% larger than the minimum value, but with $f_{Uave} = 0.75MHz$, the deviation is only 67%. On the other hand, if f_{Umax} and f_{Umin} were $0.76MHz$ and $0.74MHz$, respectively, whether their difference is divided by f_{Umin}, f_{Uave}, or f_{Umax} will not make much difference in the deviation that is calculated. So even with this inconsistent way of determining the % deviation, opamps 1a and

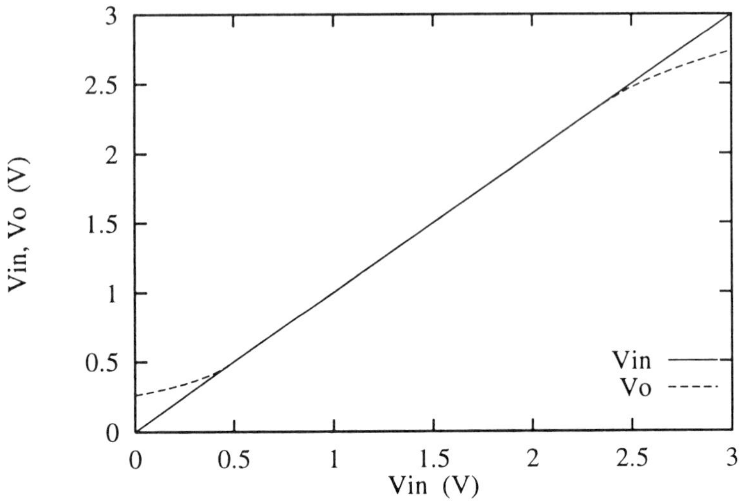

(a)

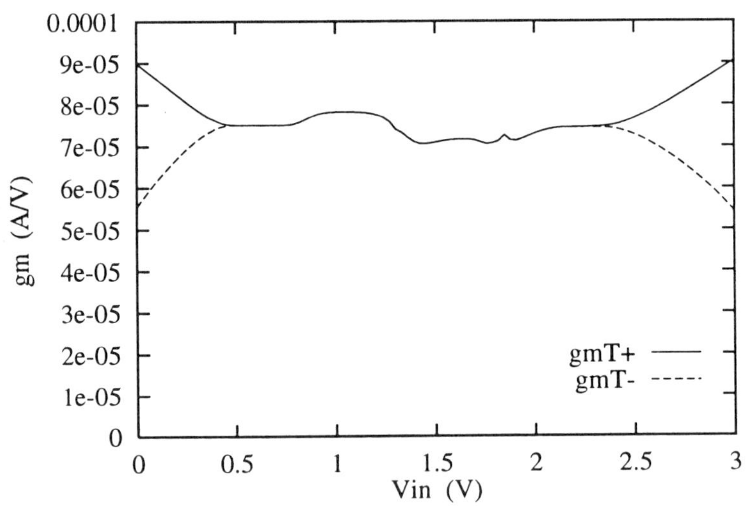

(b)

Figure 7.10: Simulation results of opamp 1b in a unity gain configuration: (a) Vin-Vo characteristics, (b) Input stage transconductance.

Table 7.11: Simulated frequency response of opamp 1b.

		$C_L = 5pF$		$C_L = 30pF$	
$V_{CM}(V)$	$A_{DC}(dB)$	$f_U(MHz)$	$\phi_M(^O)$	$f_U(MHz)$	$\phi_M(^O)$
0.5	41	1.95	87	0.382	90
0.6	56	2.22	86	0.392	89
0.7	69	2.24	86	0.393	89
0.8	69	2.26	86	0.395	89
0.9	70	2.32	86	0.406	89
1.0	70	2.34	86	0.410	89
1.1	69	2.34	86	0.410	89
1.2	69	2.32	85	0.411	89
1.3	69	2.21	84	0.390	89
1.4	69	2.09	84	0.368	89
1.5	69	2.04	83	0.372	88
1.6	69	2.02	82	0.375	88
1.7	69	1.99	83	0.372	88
1.8	70	2.04	85	0.365	88
1.9	70	2.13	82	0.372	89
2.0	70	2.16	82	0.379	89
2.1	70	2.20	82	0.386	88
2.2	69	2.21	83	0.389	88
2.3	33	1.94	85	0.378	90
2.4	16	1.56	95	0.358	98
2.5	8	1.38	110	0.325	113

1b have much more constant f_U as a function of V_{CM} when compared to opamp 1. It was shown that using the constant-g_m input stage results in a better and more uniform opamp performance throughout the linear operation range. However, that linear range is limited due to the use of the cascode stage as the output stage. The new output stage developed in the previous chapter will be used in the design of opamps and their performances are evaluated in the next chapter.

Table 7.12: Simulated common mode rejection ratio of opamp 1b.

		$1/CMRR(dB)$ @(Hz)					
$V_{CM}(V)$	$C_L(pF)$	DC	$1K$	$10K$	$100K$	$1M$	$10M$
1.0	5	101	90	70	51	31	23
1.5	5	82	78	60	40	22	22
2.0	5	84	83	68	48	29	21
1.0	30	101	90	70	51	39	38
1.5	30	82	78	60	41	30	37
2.0	30	84	83	68	48	37	36

Table 7.13: Simulated power supply rejection ratio of opamp 1b.

		$1/PSRR(dB)$ from V_{DD} @(Hz)					
$V_{CM}(V)$	$C_L(pF)$	DC	$1K$	$10K$	$100K$	$1M$	$10M$
1.0	5	62	61	59	42	22	14
1.5	5	83	79	61	41	21	14
2.0	5	62	62	57	39	19	13
1.0	30	62	61	59	42	30	29
1.5	30	83	79	61	41	29	29
2.0	30	62	62	57	39	28	28
		$1/PSRR(dB)$ from V_{SS} @(Hz)					
$V_{CM}(V)$	$C_L(pF)$	DC	$1K$	$10K$	$100K$	$1M$	$10M$
1.0	5	69	69	66	49	32	25
1.5	5	69	69	66	48	31	23
2.0	5	74	73	66	47	28	20
1.0	30	69	69	66	49	40	40
1.5	30	69	69	66	49	39	38
2.0	30	74	73	66	49	36	35

Table 7.14: Simulation results of 2% settling time of opamp 1b with $C_L = 5pF$.

$V_{in\ STEP}(V)$	$V_{in\ INIT}(V)$	$T_{S\uparrow}(\mu sec)$	$T_{S\downarrow}(\mu sec)$
0.2	0.9	0.29	0.28
0.2	1.4	0.35	0.31
0.2	1.9	0.29	0.28
0.4	1.3	0.37	0.31
0.6	1.2	0.41	0.31
0.8	1.1	0.42	0.32
1.0	1.0	0.44	0.35
1.2	0.9	0.45	0.39
1.4	0.8	0.48	0.43
1.6	0.7	0.50	0.44

Table 7.15: Simulated total harmonic distortion of opamp 1b with $C_L = 5pF$.

$f_O(Hz)$	$V_{CM}(V)$	$v_{in}(V)$	$THD(\%)$
10K	1.0	0.2	0.006
10K	1.5	0.2	0.016
10K	2.0	0.2	0.012
10K	1.5	0.5	0.039
10K	1.5	0.6	0.042
10K	1.5	0.7	0.047
10K	1.5	0.8	0.055
10K	1.5	0.9	0.21
10K	1.5	1.0	0.91
100K	1.0	0.2	0.053
100K	1.5	0.2	0.17
100K	2.0	0.2	0.12
100K	1.5	0.5	0.42
200K	1.0	0.2	0.10
200K	1.5	0.2	0.39
200K	2.0	0.2	0.24
200K	1.5	0.5	0.96

Table 7.16: Deviations in the unity gain frequency of the single stage opamps with V_{CM} varied between 0.7 and 2.2V.

	opamp 1	opamp 1a	opamp 1b
$C_L(pF) = 5pF$, $f_{U max}(MHz)$	2.70	2.37	2.34
$f_{U min}(MHz)$	1.59	2.04	1.99
$f_{U ave}(MHz)$	2.22	2.26	2.18
$\%\ deviation$	50	14.6	16
$C_L(pF) = 30pF$, $f_{U max}(MHz)$	0.479	0.410	0.410
$f_{U min}(MHz)$	0.275	0.371	0.365
$f_{U ave}(MHz)$	0.419	0.395	0.387
$\%\ deviation$	48	10	12

Chapter 8

Two-Stage Operational Amplifiers

In this chapter, we will discuss the design of two-stage opamps which have rail-to-rail input common mode range and rail-to-rail output range. These opamps are designed with the newly developed output stage shown in Figure 6.9 used with each of the three single-stage opamps designed in the previous chapter. However, in order to keep the current in the cascode stage constant, additional circuitries are added as an interface between the input and the output stage. Both single-ended and fully-differential output architectures are developed.

8.1 Single-ended Outputs

8.1.1 Opamp 2: Folded-Cascode Opamp With Rail-to-Rail Input and Output Stage

Figure 8.1 shows the architecture of opamp 2. The bottom part of the circuit diagram is the output stage shown in Figure 6.9 and the input stage is shown at the top part of the figure. As in opamp 1, M_p provides bias current to the p-channel differential pair consisting of M_1 and M_{1a}, while M_8 provides the bias current to the n-channel differential pair consisting of M_2 and M_{2a}. Note that in the design of the output stage, the only currents entering or leaving the cascode stage are small signal currents. That is, the common mode current change at the input stage must not be seen by the cascode stage. M_{1b} and M_{1c} connected in parallel with the p-channel pair carry the current which is exactly the same as the common mode current in

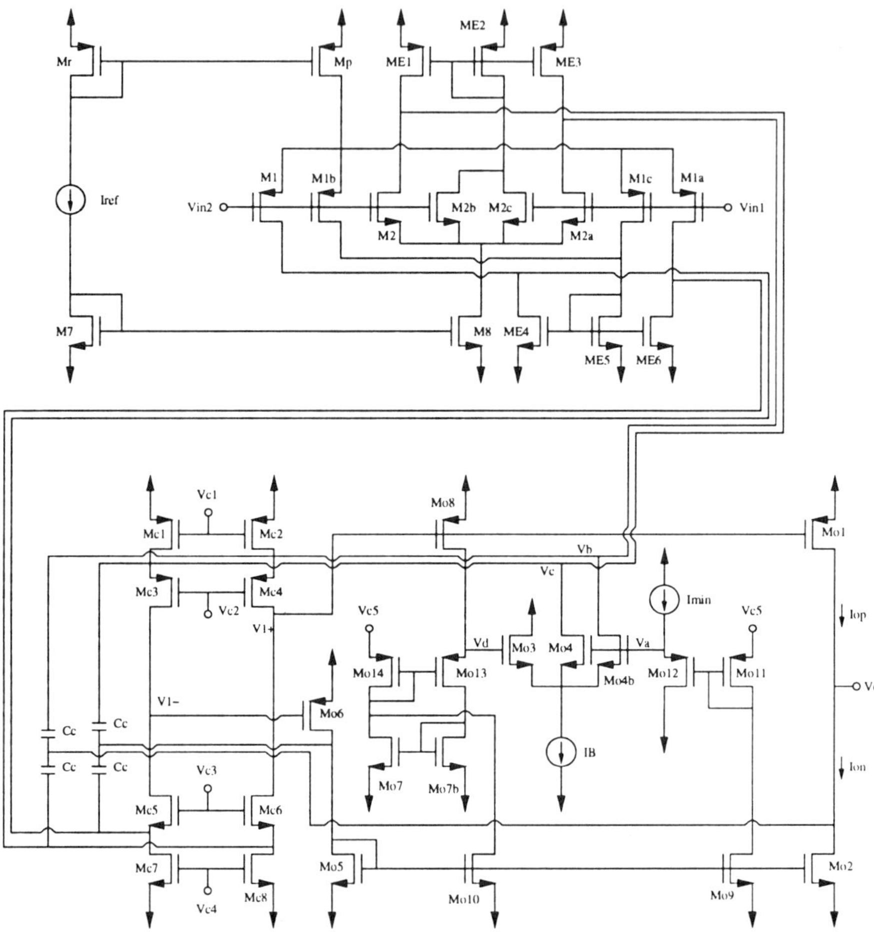

Figure 8.1: Opamp 2: A two-stage opamp with rail-to-rail input and output ranges.

the p-channel pair. Using the current mirrors consisting of M_{E4}, M_{E5}, and M_{E6}, the common mode currents in M_1 and M_{1a} are subtracted and only the small signal currents will enter the cascode stage. Transistors M_{2b}, M_{2c}, M_{E1}, M_{E2}, and M_{E3} are used in the same manner to remove the common mode current supplied from the n-channel differential pair.

Figure 8.2 shows the dc characteristics of opamp 2. In a unity gain configuration, Figure 8.2(a) shows that the output follows the input to within about $0.1V$ of the power supplies($V_{DD} = 3V$ and $V_{SS} = 0$). The output terminal was loaded with a $10K\Omega$ resistor. Figure 8.2(b) shows the input stage transconductance as a function of the common mode input voltage. Because of the rail-to-rail output stage, g_{mT} is flat below $V_{in} = 1V$ and above $V_{in} = 2V$.

Figure 8.3 shows the small signal frequency response of opamp 2. As expected, because of the larger g_{mT} for $V_{CM} = 1.5V$ compared to when $V_{CM} = 0.3V$ or $2.7V$, the low frequency gain as well as the unity gain frequency are accordingly larger for $V_{CM} = 1.5V$. Because of the two-stage design, the open loop phase of the opamp will experience a 180° phase shift at a frequency which is close to the second pole, thus we need to be concerned with the opamp gain margin, GM, as well as the phase margin, ϕ_M. Table 8.1 shows the detailed results of the open loop frequency response of opamp 2. The frequency, f_{GM}, is the frequency at which the gain margin is determined, *i.e.*, the frequency at which the phase shift is 180°. We notice that the dc gain of opamp 2 is largest when V_{CM} is about $1.5V$ and gradually decreases as V_{CM} is changed toward the supplies until it reaches $0.2V$ and $2.8V$ at which points the gain drops more sharply. Also, we see that the effect of the larger load capacitance on the frequency response is significantly less than that observed in single-stage opamps. This, of course, is due to the buffering of the high resistance cascode output node from the external loading.

Table 8.2 shows the simulation results of the common mode and power supply rejections. Because the two-stage opamp given in this section has a larger linear output range than the single-stage opamps, the rejection ratios are also measured at $V_{CM} = 0.5V$ and $2.5V$. The explanations given for the simulation results of $CMRR$ of opamp 1 apply here with two exceptions. First, the transistors added to keep the cascode stage bias current constant by subtracting the common mode current help to better reject the common mode input signal in exactly the same way. So $CMRR$ of opamp 2 is expected to be better than that of opamp 1. Second, in opamp 1, when $V_{CM} = 1.0V$ or $2.0V$, the rejection degraded due to one of the current sources entering the triode region. This still applies here; however, when V_{CM} is further reduced toward V_{SS} or increased toward V_{DD}, the differential pair which is pushing the current source into the triode region

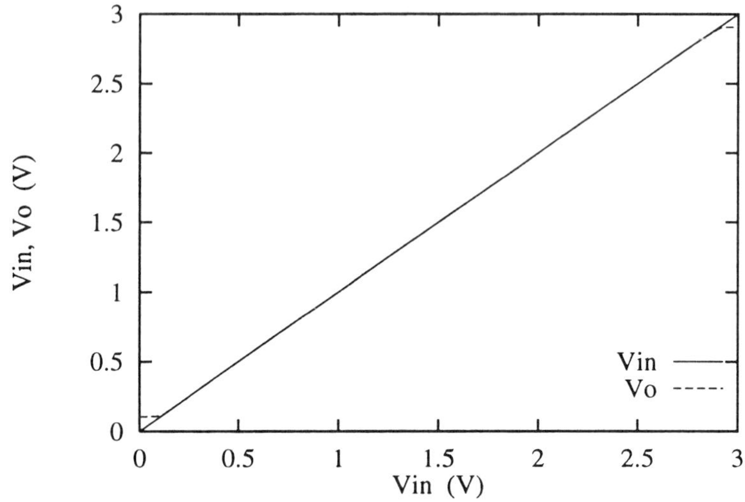

(a)

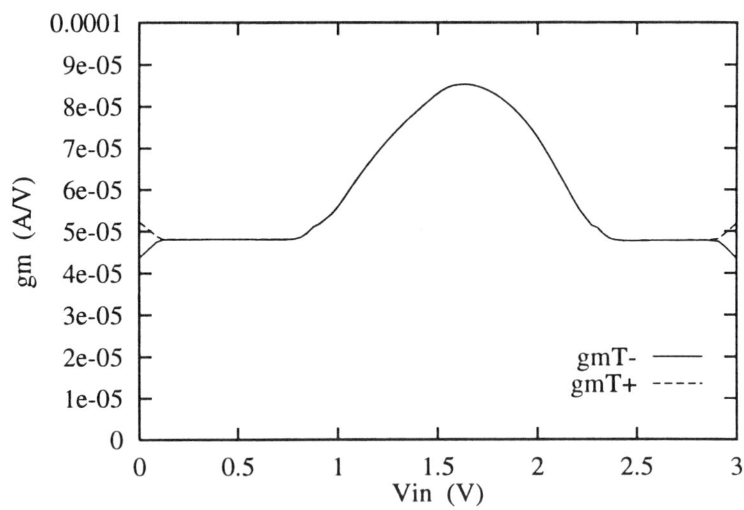

(b)

Figure 8.2: Simulation results of opamp 2 in a unity gain configuration: (a) Vin-Vo characteristics, (b) Input stage transconductance.

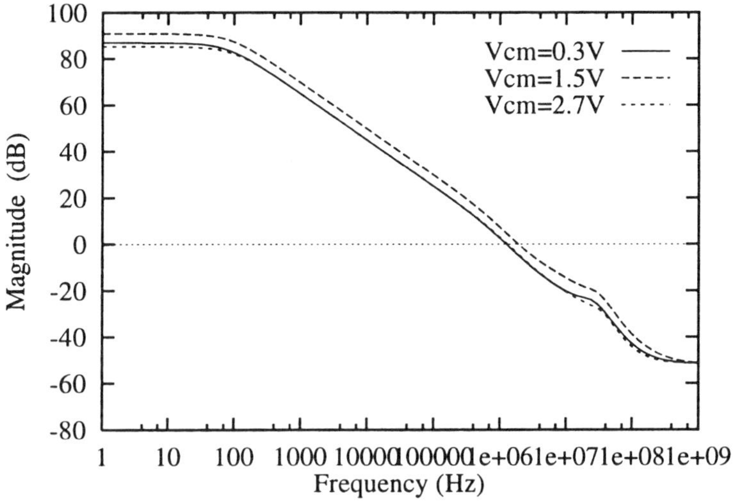

(a)

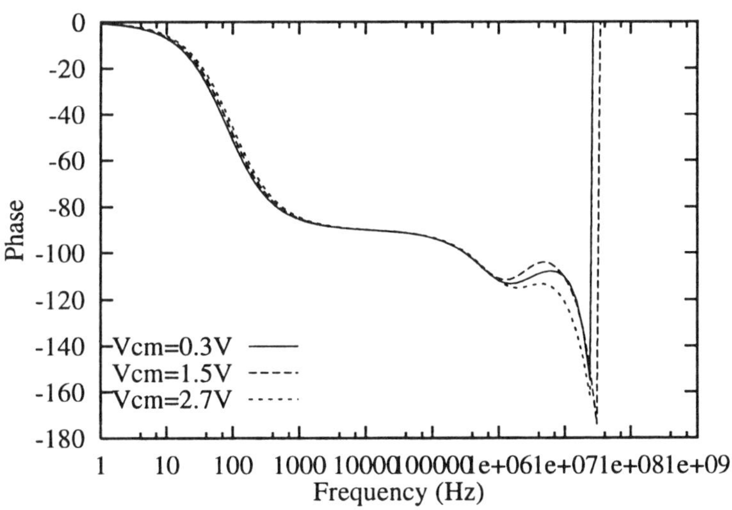

(b)

Figure 8.3: Open loop frequency response simulation of opamp 2: (a) Magnitude response, (b) Phase response.

Table 8.1: Simulated frequency response of opamp 2 with $R_L = 100K\Omega$.

		$f_U(MHz)$	$\phi_M(^O)$	$GM(dB)$	$f_{GM}(MHz)$				
$V_{CM}(V)$	$A_{DC}(dB)$	$C_L = 5pF$				$C_L = 30pF$			
0.1	58	0.62	91	28	31	0.57	79	33	19
0.2	81	1.22	68	26	30	1.01	54	32	19
0.3	87	1.25	67	26	30	1.02	53	32	19
0.4	87	1.25	67	26	30	1.02	53	32	19
0.5	87	1.25	67	26	30	1.02	53	32	19
0.6	87	1.25	67	26	30	1.02	53	33	19
0.7	87	1.25	67	26	30	1.02	54	32	19
0.8	87	1.25	67	26	30	1.03	54	32	19
0.9	87	1.32	69	25	30	1.08	55	30	19
1.0	88	1.38	69	25	30	1.14	55	30	19
1.1	89	1.50	70	24	31	1.23	55	29	19
1.2	90	1.62	70	23	31	1.33	55	28	19
1.3	90	1.71	70	22	31	1.42	56	27	19
1.4	91	1.80	70	22	31	1.48	56	27	19
1.5	91	1.88	70	21	32	1.55	55	26	19
1.6	91	1.88	70	21	32	1.60	56	26	19
1.7	91	1.88	70	21	32	1.60	55	26	19
1.8	90	1.87	69	21	32	1.58	55	26	19
1.9	90	1.82	69	22	32	1.53	55	26	19
2.0	89	1.70	68	22	32	1.45	55	27	19
2.1	88	1.56	68	24	31	1.37	56	27	18
2.2	87	1.42	67	25	31	1.25	55	29	17
2.3	86	1.34	67	26	30	1.18	56	30	16
2.4	85	1.29	66	27	29	1.14	55	30	15
2.5	85	1.29	66	27	29	1.14	55	30	15
2.6	85	1.29	66	27	29	1.14	55	30	15
2.7	85	1.29	66	27	29	1.14	55	30	15
2.8	70	1.20	70	28	25	1.08	58	30	13
2.9	55	0.87	79	29	19	0.83	68	31	10

will turn off, and there will be no propagation of the common mode signal from that side of the input stage. Then $CMRR$ will again be as good as for the case when $V_{CM} = 1.5V$.

The power supply rejection ratio is again determined by the signals entering the supply connections at the cascode stage. Unlike the case in opamp 1, $PSRR$ from the negative power supply is sensitive to the matching between transistors M_{c7} and M_{c8} because of the absence of the local feedback between V_{1-} and V_{c4} terminals. Since the matching between the small signal resistance at V_{1+} and V_{1-} affects $PSRR$ from either supplies the best $PSRR$ is obtained when V_{1+} is approximately the same as V_{1-}. This occurs when the current flowing through M_{O1} is the same as the current in M_{O6}, and that of course happens when $V_o = 1.5V$. As V_o further moves away from $1.5V$, V_{1+} and V_{1-} further differ from each other and this causes $PSRR$ to get worse as we can see in Table 8.2.

Table 8.3 shows the 2% settling time of the opamp 2 in a unity gain configuration. Because two stage opamps have at least two well defined poles, the step response will be underdamped and there will be some overshoot. The overshoot is measured as a percentage overshoot, $P.O$, and is given in Table 8.3. The settling time of opamp 2 is larger than that of opamp 1 because of the smaller unity gain frequency. As in opamp 1, for a given step size, both $T_S \uparrow$ and $T_{S\downarrow}$ are the smallest when the input step is positioned around $1.5V$ because this results in g_{mT}, and hence, f_U to be at their maximum values.

The harmonic distortion of opamp 2 determined by simulation is listed in Table 8.4. When an input sine wave with $0.5V$ amplitude is applied, THD is the smallest when the dc component of the input source is $1.5V$ as opposed to $1.0V$ or $2.0V$. This was the case for opamp 1 and the same explanation will apply here. For the same input values, THD in opamp 1 and opamp 2 are almost the same and this implies that the output stage does not cause any extra distortion to the opamp.

8.1.2 Opamp 2a: Rail-to-Rail Folded-Cascode Opamp With Constant-g_m Input Stage 1

Figure 8.4 shows the opamp with rail-to-rail input and output stage and with the constant-g_m circuit 1. This circuit is the two-stage version of opamp 1a. The currents in the cascode stage are maintained constant by $M_{E1} - M_{E6}$. The common mode current in the p-channel differential pair, that is dependent on the common mode input voltage, is already available in M_3. Adding M_{E4} and M_{E5} to subtract the current from M_1 and M_{1a} will eliminate the current change in the cascode stage caused by the p-channel input pair. For the common mode current in the n-channel input pair, I_n is

Table 8.2: Simulated common mode and power supply rejection ratio of opamp 2 with $R_L = 100K\Omega$.

		$1/CMRR(dB)$	$@(Hz)$				
$V_{CM}(V)$	$C_L(pF)$	DC	$1K$	$10K$	$100K$	$1M$	$10M$
0.5	5	118	106	86	66	45	34
1.0	5	96	92	74	53	24	27
1.5	5	119	95	75	55	27	22
2.0	5	93	92	80	60	28	21
2.5	5	123	104	83	63	41	33
0.5	30	118	106	86	66	44	39
1.0	30	96	92	74	53	22	32
1.5	30	119	95	75	55	26	27
2.0	30	93	92	80	60	27	25
2.5	30	123	104	83	63	40	38
		$1/PSRR(dB)$ from V_{DD}					
0.5	5	66	51	32	12	-7	7
1.0	5	74	53	33	13	-7	6
1.5	5	90	56	36	16	-3	6
2.0	5	76	55	35	15	-4	5
2.5	5	67	52	32	12	-8	5
0.5	30	66	52	32	12	-8	12
1.0	30	74	53	33	13	-8	12
1.5	30	90	56	36	16	-4	10
2.0	30	76	55	35	15	-5	9
2.5	30	67	52	32	12	-9	10
		$1/PSRR(dB)$ from V_{SS}					
0.5	5	65	49	29	9	-10	3
1.0	5	72	50	30	10	-9	3
1.5	5	95	54	34	14	-6	2
2.0	5	80	53	33	13	-7	2
2.5	5	71	49	29	9	-10	2
0.5	30	65	49	29	9	-11	8
1.0	30	72	50	30	10	-11	8
1.5	30	95	54	34	13	-7	7
2.0	30	80	53	33	14	-8	6
2.5	30	71	49	29	9	-11	6

Table 8.3: Simulation results of 2% settling time of opamp 2 with $R_L = 100K\Omega$ and $C_L = 5pF$.

$V_{in\,STEP}(V)$	$V_{in\,INIT}(V)$	$T_{S\uparrow}(\mu sec)$	$T_{S\downarrow}(\mu sec)$	$P.O_{\uparrow}(\%)$	$P.O_{\downarrow}(\%)$
0.4	0.6	0.62	0.64	10	7.5
0.4	1.3	0.46	0.45	10	10
0.4	2.0	0.58	0.50	7.5	7.5
1.0	1.0	0.59	0.55	9.0	8.0
1.6	0.7	0.71	0.63	7.5	5.6
2.4	0.3	0.84	0.75	5.4	4.6

available at M_7 and hence the current is mirrored into the drain terminals of M_2 and M_{2a} using M_{E6}, M_{E3}, M_{E2}, and M_{E1}.

Figure 8.5 shows the dc characteristics of opamp 2a. Figure 8.5(a) shows that the output follows the input to within about $0.1V$ of the power supplies($V_{DD} = 3V$ and $V_{SS} = 0$) as was the case for opamp 2. Again, the output terminal was loaded with a $10K\Omega$ resistor. Figure 8.5(b) shows the input stage transconductance as a function of the common mode input voltage. Because of the rail-to-rail output stage, together with the constant-g_m input stage, g_{mT} is fairly constant almost from rail-to-rail.

The small signal frequency response of opamp 2a is shown in Figure 8.6; unlike the case in opamp 2, we see that the unity gain frequency of the opamp 2a is almost the same for $V_{CM} = 0.3V$, $1.5V$, and $2.7V$. Table 8.5 shows more detailed results of the frequency response and we can clearly see the effectiveness of the constant-g_m input stage in keeping f_U constant.

The common mode and the power supply rejection ratios of opamp 2a are listed in Table 8.6. Again the worst $CMRR$ is obtained when $V_{CM} = 2.0V$ and the current source supplying the p-channel differential pair goes into the triode region. Since the power supply rejection ratio is again determined by the signal entering the cascode stage, and since we have made efforts to keep the operating condition of the cascode stage to be unaffected by the input common mode voltage, $PSRR$ from both supplies should behave in a similar manner to that in opamp 2. Indeed, we see that the rejection is the best when $V_{CM} = 1.5V$ and starts to degrade as V_{CM} is changed toward the supplies. Note that this would be true for opamp 2b as well.

Table 8.7 shows the 2% settling time of opamp 2a in a unity gain configuration. Compared to the step response of opamp 2, for $0.4V$ input step, the trend seems to show a slower settling for a step near $1.5V$ and faster

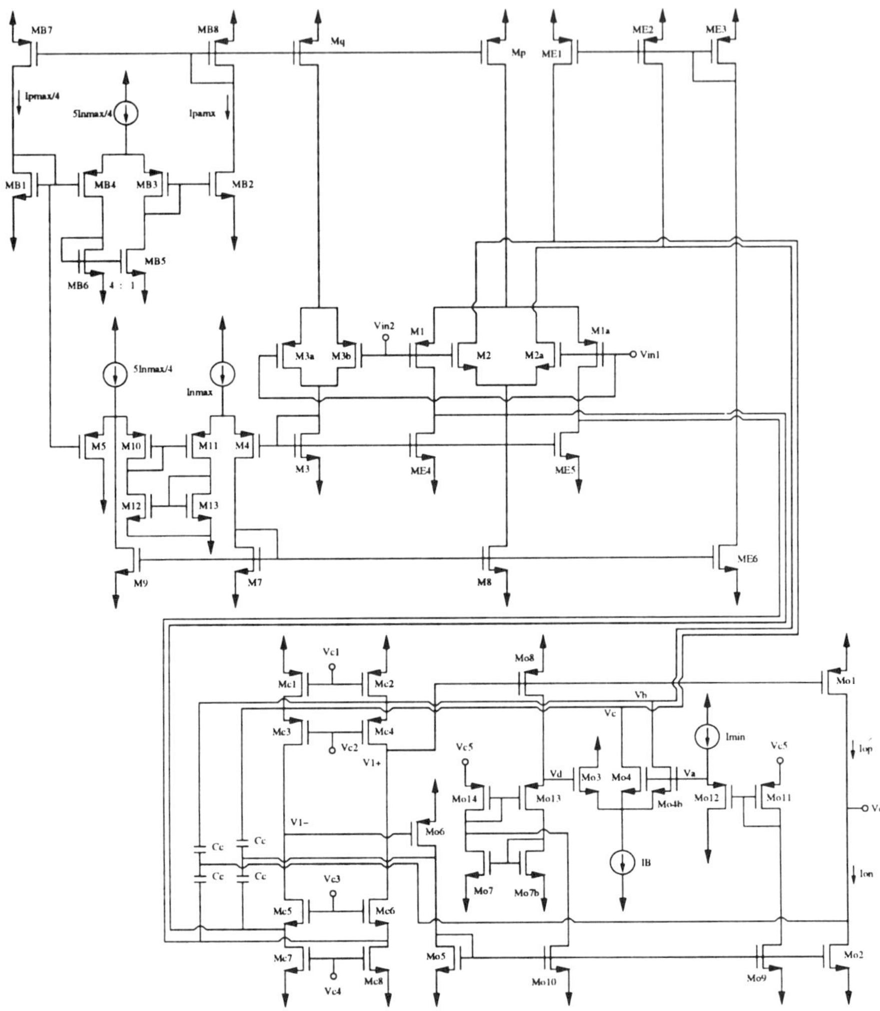

Figure 8.4: Opamp 2a: A rail-to-rail two-stage opamp with the constant-g_m input stage 1.

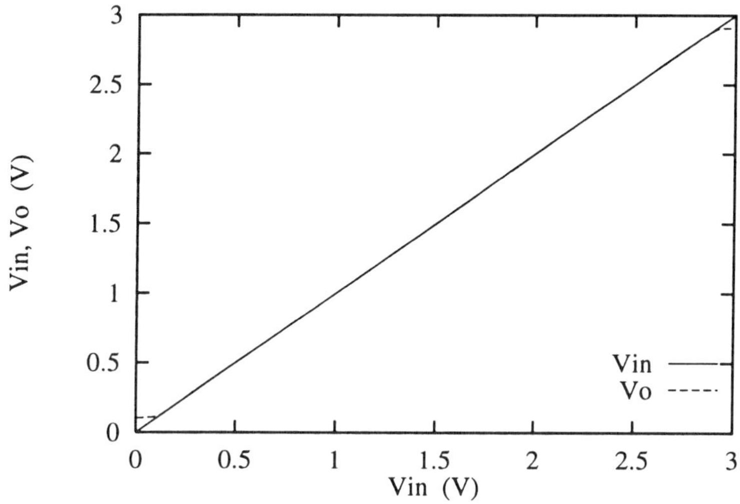

(a)

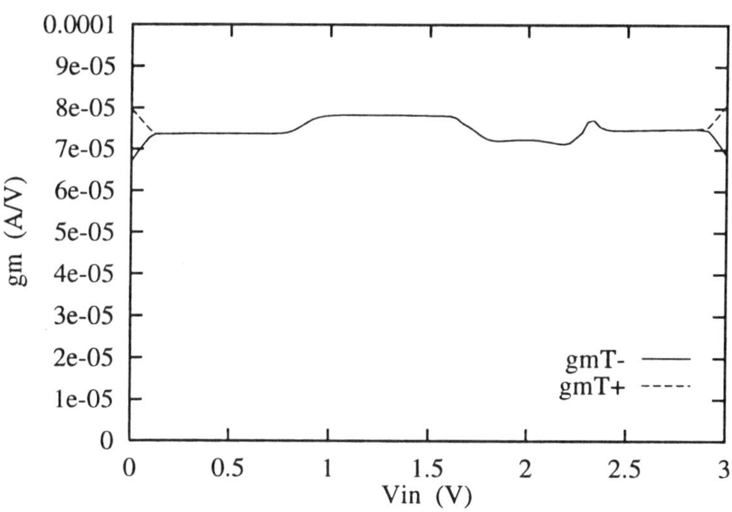

(b)

Figure 8.5: Simulation results of opamp 2a in a unity gain configuration: (a) Vin-Vo characteristics, (b) Input stage transconductance.

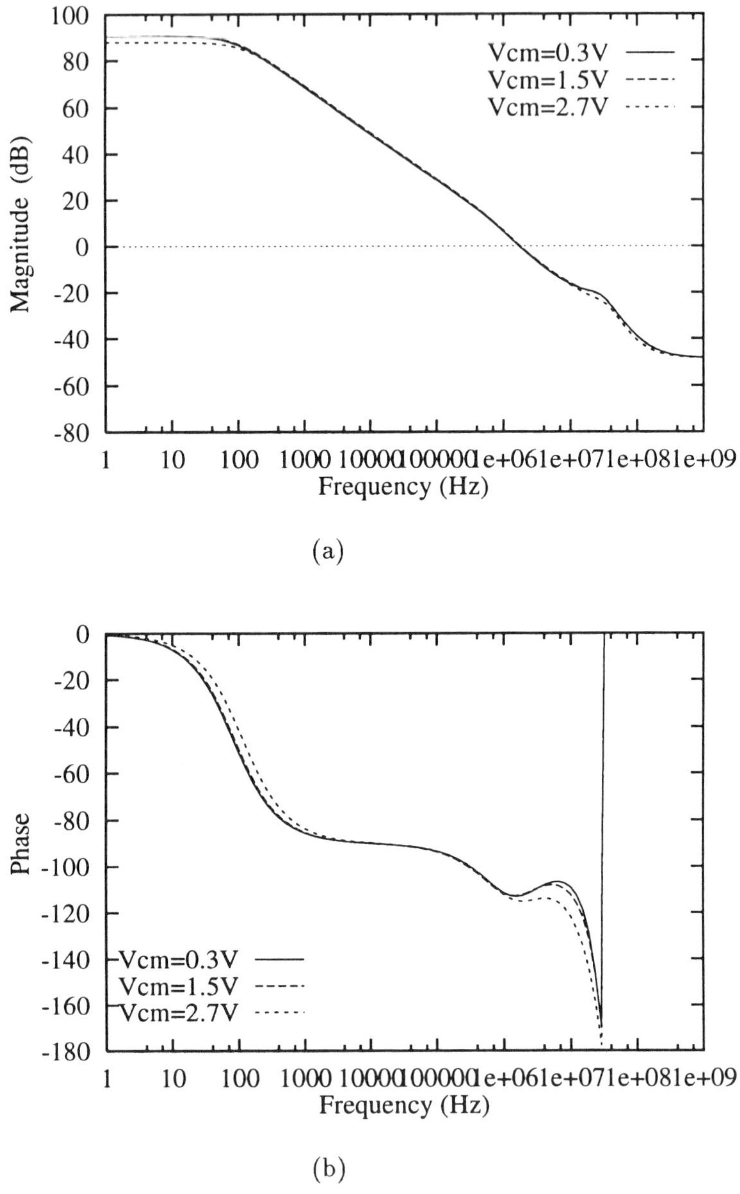

Figure 8.6: Open loop frequency response simulation of opamp 2a: (a) Magnitude response, (b) Phase response.

Table 8.4: Simulated total harmonic distortion of opamp 2 with $R_L = 100K\Omega$.

$C_L(pF)$	$f_O(Hz)$	$V_{CM}(V)$	$v_{in}(V)$	$THD(\%)$
5	10K	1.0	0.5	0.088
5	10K	1.5	0.5	0.037
5	10K	2.0	0.5	0.092
5	10K	1.5	1.0	0.096
5	10K	1.5	1.1	0.098
5	10K	1.5	1.2	0.10
5	10K	1.5	1.3	0.11
5	10K	1.5	1.4	0.12
5	100K	1.0	0.5	0.88
5	100K	1.5	0.5	0.38
5	100K	2.0	0.5	0.92
5	100K	1.5	1.0	0.95
5	200K	1.0	0.5	2.0
5	200K	1.5	0.5	0.80
5	200K	2.0	0.5	1.9
5	200K	1.5	1.0	2.1
30	10K	1.0	0.5	0.092
30	10K	1.5	0.5	0.037
30	10K	2.0	0.5	0.088
30	100K	1.0	0.5	0.94
30	100K	1.5	0.5	0.39
30	100K	2.0	0.5	0.91
30	200K	1.0	0.5	2.0
30	200K	1.5	0.5	0.92
30	200K	2.0	0.5	2.2

settlings for steps near 0.8 and $2.2V$ possibly due to more constant unity gain frequency of opamp 2a as a function of the common mode input voltage. For larger input steps, $T_{S\downarrow}$ is significantly less than $T_{S\uparrow}$. The slew rate limitation is related to the speed and the amount of voltage changes at the V_{1+} and V_{1-} nodes. First, consider the case when a large positive step is applied. When V_{in1} and V_{in2} are close to V_{SS}, currents I_1, I_{1a}, I_{E4}, and I_{E5} are large and they should all be the same. When V_{in1} is suddenly moved toward V_{DD}, I_1 increases and I_{1a} decreases. At the first part of the settling behavior, V_{in2} is still low enough to keep M_{3a} conducting current. Then I_{E4} and I_{E5} should stay more or less constant. In order for V_o to

Table 8.5: Simulated frequency response of opamp 2a with $R_L = 100K\Omega$.

		$f_U(MHz)$	$\phi_M(^O)$	$GM(dB)$	$f_{GM}(MHz)$				
$V_{CM}(V)$	$A_{DC}(dB)$	$C_L = 5pF$				$C_L = 30pF$			
0.1	59	0.86	82	25	32	0.76	79	29	19
0.2	79	1.57	68	23	31	1.26	55	28	19
0.3	89	1.66	67	23	31	1.33	52	28	19
0.4	89	1.66	67	23	31	1.33	52	28	19
0.5	89	1.65	67	23	31	1.33	53	28	19
0.6	89	1.65	67	23	31	1.33	53	28	19
0.7	89	1.65	67	23	31	1.33	53	28	19
0.8	89	1.66	67	23	31	1.34	53	28	19
0.9	89	1.72	67	23	31	1.39	53	28	19
1.0	89	1.74	67	22	31	1.41	53	28	19
1.1	89	1.74	67	22	31	1.41	53	28	19
1.2	89	1.74	67	22	31	1.42	53	28	19
1.3	89	1.73	67	22	31	1.42	53	28	19
1.4	89	1.73	67	22	31	1.42	53	28	19
1.5	89	1.73	67	22	31	1.42	53	29	19
1.6	89	1.73	67	23	31	1.43	53	29	19
1.7	89	1.68	66	22	31	1.39	54	28	19
1.8	88	1.65	68	22	31	1.38	56	28	20
1.9	88	1.66	71	22	31	1.37	58	28	19
2.0	88	1.65	74	22	31	1.37	60	27	18
2.1	88	1.69	74	23	30	1.40	61	27	17
2.2	87	1.74	71	23	29	1.45	59	27	16
2.3	87	1.80	67	23	29	1.53	54	27	16
2.4	87	1.75	65	24	29	1.50	53	27	16
2.5	87	1.75	64	24	29	1.51	52	27	15
2.6	87	1.75	63	24	29	1.52	52	27	15
2.7	87	1.76	63	24	29	1.53	52	27	15
2.8	72	1.69	66	24	26	1.46	53	27	14
2.9	57	1.23	74	24	17	1.18	60	27	10

Table 8.6: Simulated common mode and power supply rejection ratio of opamp 2a with $R_L = 100K\Omega$.

		$1/CMRR(dB)$ @(Hz)					
$V_{CM}(V)$	$C_L(pF)$	DC	$1K$	$10K$	$100K$	$1M$	$10M$
0.5	5	126	123	107	85	54	32
1.0	5	123	112	92	72	53	31
1.5	5	149	132	113	92	68	30
2.0	5	103	91	71	50	17	20
2.5	5	147	113	93	72	44	31
0.5	30	126	123	107	85	53	36
1.0	30	123	112	92	72	52	36
1.5	30	149	132	113	92	67	35
2.0	30	103	91	71	50	16	25
2.5	30	147	113	93	72	43	35
		$1/PSRR(dB)$ from V_{DD}					
0.5	5	71	56	36	16	-4	6
1.0	5	78	56	36	16	-3	6
1.5	5	86	56	36	16	-4	6
2.0	5	73	55	35	15	-4	5
2.5	5	69	55	36	16	-4	5
0.5	30	71	56	36	16	-5	11
1.0	30	78	56	36	16	-5	11
1.5	30	86	56	36	16	-5	11
2.0	30	73	55	35	15	-5	10
2.5	30	69	55	36	16	-5	10
		$1/PSRR(dB)$ from V_{SS}					
0.5	5	70	53	33	13	-6	3
1.0	5	76	53	33	13	-7	3
1.5	5	90	53	33	13	-7	3
2.0	5	75	52	32	12	-6	2
2.5	5	72	53	33	13	-7	2
0.5	30	70	53	33	13	-8	8
1.0	30	76	53	33	13	-8	8
1.5	30	90	53	33	13	-8	8
2.0	30	75	52	32	12	-7	8
2.5	30	72	53	33	13	-8	7

Table 8.7: Simulation results of 2% settling time of opamp 2a with $R_L = 100K\Omega$ and $C_L = 5pF$.

$V_{in\ STEP}(V)$	$V_{in\ INIT}(V)$	$T_{S\uparrow}(\mu sec)$	$T_{S\downarrow}(\mu sec)$	$P.O_{\uparrow}(\%)$	$P.O_{\downarrow}(\%)$
0.4	0.6	0.57	0.51	10	10
0.4	1.3	0.59	0.49	10	10
0.4	2.0	0.49	0.65	7.5	18
1.0	1.0	0.64	0.50	9.0	8.0
1.6	0.7	0.73	0.54	7.5	5.6
2.4	0.3	0.89	0.66	5.4	4.6

move toward V_{DD}, V_{1+} must be reduced first, and it moves as quickly as the drain terminal of M_{1a} does; the rate of change at this node is limited by the capacitance at this node(C_C) and the current available to charge it up. This current is the difference in I_{E5} and I_{1a}; and since I_{E5} is almost constant at first, the change in I_{1a} will affect $T_{S\uparrow}$. For a large negative step input, the rate of change in V_o moving toward V_{SS} is determined by the rate of change in V_{1-} and hence the drain terminal of M_2 moves toward V_{SS}. This will be determined by the amount of current, $I_2 - I_{E1}$, which is available to charge up C_C connected to this node. When V_{in1} is suddenly moved toward V_{SS}, I_2 will increase; also, M_{3a}, which was turned off, will be on and sink some current into M_3. Then, this will reduce I_{E1} and I_{E2}. This implies that $I_2 - I_{E1}$ can be quite large because I_2 will be increasing while I_{E1} is decreasing, and this helps $T_{S\downarrow}$ to be small.

The harmonic distortion of opamp 2a is shown in Table 8.8. We see that THD of opamp 2a is significantly less than that of opamp 2, especially at low frequencies. However, we must be cautious as THD simulation results are usually not perfectly accurate. Nevertheless, the results indicate that we can expect some improvements in THD when the actual circuit is tested.

8.1.3 Opamp 2b: Rail-to-Rail Folded-Cascode Opamp With Constant-g_m Input Stage 2

Figure 8.7 shows opamp 2b which has rail-to-rail input and output range and uses the constant-g_m input stage 2. This circuit is the two-stage version of opamp 1b. The currents in the cascode stage is maintained constant by $M_{E1} - M_{E6}$ in exactly the same way used in opamp 2a.

Figure 8.8 shows the dc characteristics of opamp 2b. Figure 8.8(a) shows that the output follows the input to within about $0.1V$ of the power supplies($V_{DD} = 3V$ and $V_{SS} = 0$) as was the case in opamp 2 and opamp 2a.

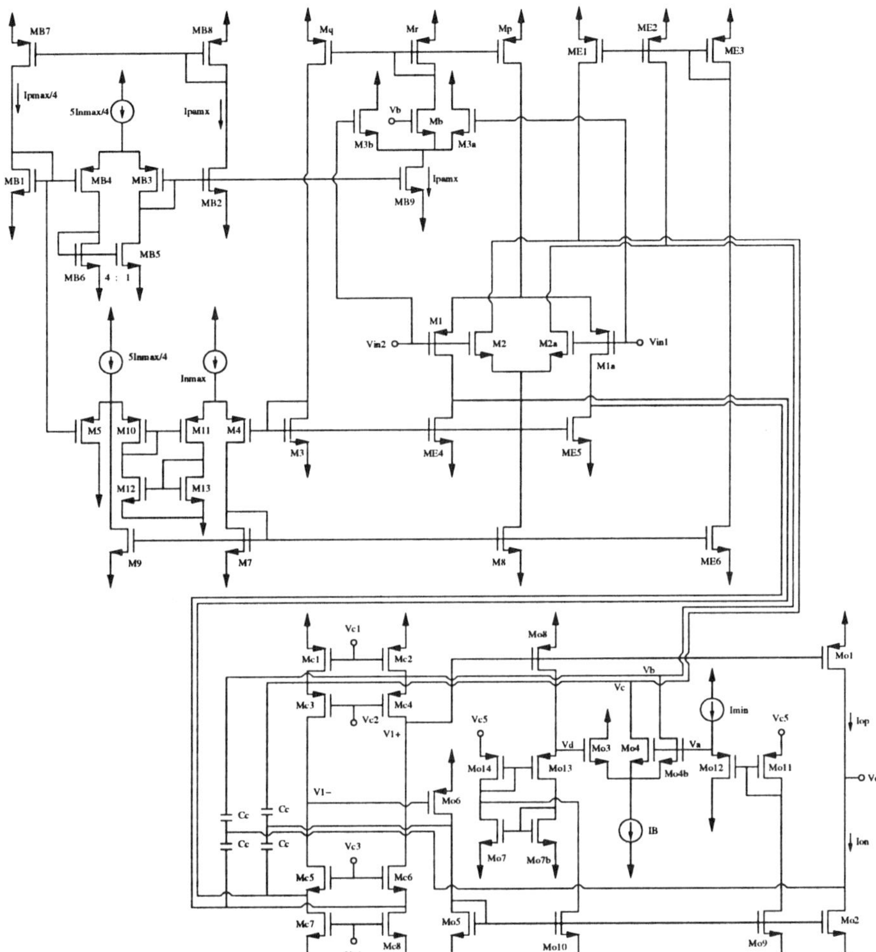

Figure 8.7: Opamp 2b: A rail-to-rail two-stage opamp with the constant-g_m input stage 2.

Table 8.8: Simulated total harmonic distortion of opamp 2a with $R_L = 100K\Omega$.

$C_L(pF)$	$f_O(Hz)$	$V_{CM}(V)$	$v_{in}(V)$	$THD(\%)$
5	10K	1.0	0.5	0.012
5	10K	1.5	0.5	0.007
5	10K	2.0	0.5	0.015
5	10K	1.5	1.0	0.014
5	10K	1.5	1.1	0.011
5	10K	1.5	1.2	0.011
5	10K	1.5	1.3	0.012
5	10K	1.5	1.4	0.028
5	100K	1.0	0.5	0.13
5	100K	1.5	0.5	0.10
5	100K	2.0	0.5	0.40
5	100K	1.5	1.0	0.32
5	200K	1.0	0.5	0.31
5	200K	1.5	0.5	0.37
5	200K	2.0	0.5	1.4
5	200K	1.5	1.0	1.2
30	10K	1.0	0.5	0.012
30	10K	1.5	0.5	0.007
30	10K	2.0	0.5	0.015
30	100K	1.0	0.5	0.15
30	100K	1.5	0.5	0.10
30	100K	2.0	0.5	0.39
30	200K	1.0	0.5	0.42
30	200K	1.5	0.5	0.38
30	200K	2.0	0.5	1.4

Again, the output terminal was loaded with a $10K\Omega$ resistor. Figure 8.5(b) shows the input stage transconductance as a function of the common mode input voltage. Just as with opamp 2a, opamp 2b has g_{mT} that is fairly constant almost from rail-to-rail.

Table 8.9 shows detailed results of the frequency response of opamp 2b, and we can again see the effectiveness of the constant-g_m input stage in keeping f_U constant.

The simulation results of the common mode and the power supply rejection ratio are shown in Table 8.10. Again, $CMRR$ is the worst when there is a signal path from the gate terminals of M_{3a} and M_{3b} to the drain ter-

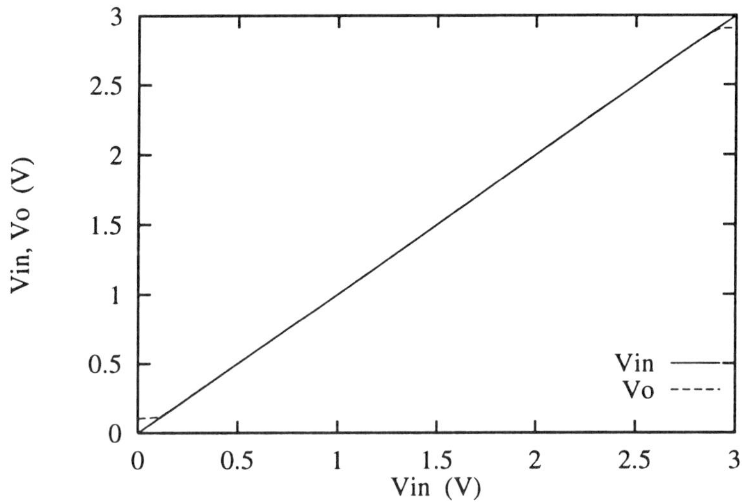

(a)

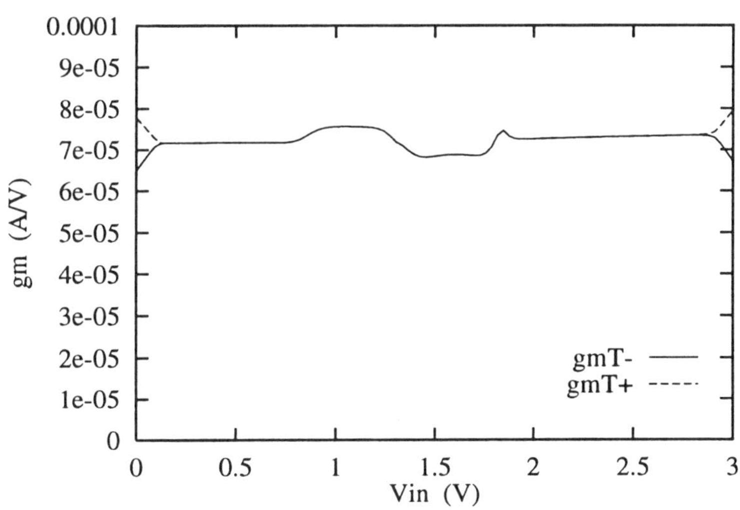

(b)

Figure 8.8: Simulation results of opamp 2b in a unity gain configuration: (a) Vin-Vo characteristics, (b) Input stage transconductance.

Table 8.9: Simulated frequency response of opamp 2b with $R_L = 100K\Omega$.

		$f_U(MHz)$	$\phi_M(^O)$	$GM(dB)$	$f_{GM}(MHz)$				
$V_{CM}(V)$	$A_{DC}(dB)$	$C_L = 5pF$				$C_L = 30pF$			
0.1	59	0.84	93	25	32	0.76	80	29	20
0.2	77	1.57	70	23	31	1.28	57	28	19
0.3	89	1.71	67	23	32	1.37	53	28	19
0.4	89	1.71	67	23	32	1.38	53	28	19
0.5	89	1.71	67	23	32	1.38	53	28	19
0.6	89	1.71	67	23	32	1.39	54	28	19
0.7	89	1.71	67	23	32	1.40	54	28	19
0.8	89	1.72	67	23	32	1.41	54	28	19
0.9	90	1.77	68	23	33	1.45	54	27	19
1.0	90	1.79	67	23	33	1.47	54	27	19
1.1	90	1.79	67	23	34	1.48	54	28	20
1.2	90	1.78	68	23	33	1.48	55	28	20
1.3	89	1.71	68	23	33	1.43	55	28	19
1.4	89	1.67	68	23	32	1.39	56	28	19
1.5	89	1.65	71	23	31	1.37	58	28	18
1.6	88	1.69	73	24	30	1.37	60	28	17
1.7	88	1.76	71	24	30	1.42	58	28	17
1.8	87	1.81	65	24	29	1.51	54	28	16
1.9	87	1.72	65	24	29	1.46	53	27	16
2.0	87	1.72	65	24	29	1.47	53	27	16
2.1	87	1.73	65	24	29	1.47	53	27	16
2.2	87	1.73	65	24	29	1.49	53	27	16
2.3	87	1.74	65	24	29	1.49	53	27	16
2.4	87	1.74	65	24	29	1.50	52	27	15
2.5	87	1.75	65	24	29	1.51	52	27	15
2.6	87	1.76	65	24	29	1.51	52	27	15
2.7	87	1.76	65	24	29	1.53	52	27	15
2.8	72	1.69	67	24	26	1.47	53	27	13
2.9	56	1.29	73	25	18	1.19	60	27	10

minal of M_r. This is the case when V_b and V_{CM} are close to each other and all M_b, M_{3a}, M_{3b} are operating. The power supply rejection ratio follows a pattern similar to that in other two stage-opamps. That is, closer values of V_{1+} and V_{1-} result in better resistance matching at those terminals and hence better cancellation of the signal entering the left and the right side of the cascode stage.

Table 8.11 shows the 2% settling time of opamp 2b in a unity gain configuration. The settling time of a $0.4V$ step around $0.8V$, $1.5V$, and $2.2V$ seems to be slightly more constant than in opamp 2a. However, the difference between $T_{S\uparrow}$ and $T_{S\downarrow}$ for large step inputs is slightly more significant than in opamp 2a. Basically the reason that $T_{S\uparrow}$ is larger than $T_{S\downarrow}$ is the same reason which was given for opamp 2a: $I_{E5} - I_{1a}$ is smaller than $I_2 - I_{E1}$.

The harmonic distortion of opamp 2b is shown in Table 8.12. The performance seems to be the way we expected it to be except for the case when the dc component of the input signal is $2.0V$. When the dc component is $1.5V$, THD for the sinusoid with amplitude all the way up to 1.4 is significantly better than that of opamp 2. Also, when the dc component is $1.0V$, THD is close to or actually better than the case for $1.5V$ dc component, at least at low frequencies. However, for a $10KHz$ sinusoid with $0.5V$ amplitude, THD of opamp 2b is actually slightly worse than that of opamp 2. Further simulations reveal that even for a very small sinusoid, when it swings around $1.9V$, the distortion is quite large. The large distortion must be caused by the discontinuity in the opamp gain around that region; and it could be a problem associated with the circuit design or it could be caused by the transistor model used in the simulation. Note that opamp 1b, which has the same input stage as opamp 2b, had good THD, and so was the THD of opamp 2a, which has the same output stage(including the cascode stage).

Simulated quiescent power dissipation of the opamps are $0.8mW$ and $1.1mW$ for opamp 2 and opamps 2a and 2b, respectively. To compare the simulation results of the two-stage opamps, the figures given in Tables 8.1, 8.5, and 8.9 are used to compare how constant f_U is as a function of the input common mode voltage. For $V_{CM} = 0.3V$ to $2.7V$, $f_{U ave}$, $f_{U max}$ and $f_{U min}$ of the opamps discussed here are shown in Table 8.13 along with the percentage deviation in f_U. The deviations are about 45% for opamp 2 and 9% to 14% for the opamps with the constant-g_m input stage.

Table 8.10: Simulated common mode and power supply rejection ratio of opamp 2b with $R_L = 100K\Omega$.

		$1/CMRR(dB)$ $@(Hz)$					
$V_{CM}(V)$	$C_L(pF)$	DC	$1K$	$10K$	$100K$	$1M$	$10M$
0.5	5	117	99	79	59	39	33
1.0	5	123	98	78	58	38	33
1.5	5	129	90	70	49	19	28
2.0	5	127	112	93	73	45	35
2.5	5	130	113	93	73	45	34
0.5	30	117	99	79	59	38	38
1.0	30	123	98	78	58	37	37
1.5	30	129	90	70	49	18	33
2.0	30	127	112	93	73	44	39
2.5	30	130	113	93	73	44	39
		$1/PSRR(dB)$ from V_{DD}					
0.5	5	71	70	35	15	-5	5
1.0	5	82	55	35	15	-5	5
1.5	5	87	54	34	14	-6	5
2.0	5	72	54	34	14	-6	5
2.5	5	67	54	34	14	-6	5
0.5	30	71	70	35	15	-6	10
1.0	30	82	55	35	15	-6	9
1.5	30	87	54	34	14	-7	10
2.0	30	72	54	34	14	-7	9
2.5	30	67	54	34	14	-7	9
		$1/PSRR(dB)$ from V_{SS}					
0.5	5	71	51	31	11	-8	2
1.0	5	81	52	32	12	-8	1
1.5	5	92	50	30	10	-9	1
2.0	5	75	51	31	11	-9	1
2.5	5	69	51	31	11	-9	1
0.5	30	71	51	31	11	-10	6
1.0	30	81	52	32	12	-9	5
1.5	30	92	50	30	10	-11	6
2.0	30	75	51	31	11	-10	5
2.5	30	69	51	31	11	-10	5

Table 8.11: Simulation results of 2% settling time of opamp 2b with $R_L = 100K\Omega$ and $C_L = 5pF$.

$V_{in\ STEP}(V)$	$V_{in\ INIT}(V)$	$T_{S\uparrow}(\mu sec)$	$T_{S\downarrow}(\mu sec)$	$P.O_{\uparrow}(\%)$	$P.O_{\downarrow}(\%)$
0.4	0.6	0.54	0.49	10	10
0.4	1.3	0.44	0.41	5.0	7.5
0.4	2.0	0.54	0.47	13	10
1.0	1.0	0.65	0.33	6.0	3.0
1.6	0.7	0.80	0.38	5.6	1.9
2.4	0.3	0.93	0.54	4.6	2.1

8.2 Fully-Differential Outputs

In this section, two fully-differential opamps are introduced. Unlike the case in single-ended opamps, we cannot directly apply common mode input voltages to the input terminals of the fully-differential opamps. This is because of the fact that each of the output must have a negative feedback. Thus, we must use the opamp in some resistive feedback configurations. Figure 8.9 shows an inverting gain configuration with a gain of -1, *i.e.*,

$$V_{o1} - V_{o2} = -(V_{i1} - V_{i2}). \tag{8.1}$$

This circuit is used for simulation of the common mode and the power supply rejection ratio, step response, and total harmonic distortion. The open loop frequency response is simulated using the circuit shown in Figure 8.10. The inductors and the capacitors are chosen to be very large, so that, at ac, inductors in the feedbacks are open circuited and the capacitors at the input side are short circuited to allow the small signals applied at V_{i1} and V_{i2} terminals to directly appear at the input terminals of the opamp. At dc, inductors are short circuited and capacitors are open circuited, so that, the circuit is operating in the simple inverting gain mode and the operating points can be determined without any convergence problem. Resistors, R, are set large so that the only resistive loading is caused by R_L.

8.2.1 Opamp 3a: Fully-Differential Rail-to-Rail Folded - Cascode Opamp With Constant-g_m Input Stage 1

Figure 8.11 shows the fully-differential version of opamp 2a. The only difference between opamps 2a and 3a is the addition of four transistors M_{o1b}, M_{o2b}, M_{o5b}, and M_{o6b}, which provide the negative output terminal,

Table 8.12: Simulated total harmonic distortion of opamp 2b with $R_L = 100K\Omega$.

$C_L(pF)$	$f_O(Hz)$	$V_{CM}(V)$	$v_{in}(V)$	$THD(\%)$
5	10K	1.0	0.5	0.023
5	10K	1.5	0.5	0.026
5	10K	2.0	0.5	0.12
5	10K	1.5	1.0	0.025
5	10K	1.5	1.1	0.023
5	10K	1.5	1.2	0.021
5	10K	1.5	1.3	0.021
5	10K	1.5	1.4	0.035
5	100K	1.0	0.5	0.22
5	100K	1.5	0.5	0.37
5	100K	2.0	0.5	0.40
5	100K	1.5	1.0	0.40
5	200K	1.0	0.5	1.2
5	200K	1.5	0.5	0.36
5	200K	2.0	0.5	1.3
5	200K	1.5	1.0	2.4
30	10K	1.0	0.5	0.023
30	10K	1.5	0.5	0.026
30	10K	2.0	0.5	0.12
30	100K	1.0	0.5	0.22
30	100K	1.5	0.5	0.36
30	100K	2.0	0.5	1.2
30	200K	1.0	0.5	0.43
30	200K	1.5	0.5	1.2
30	200K	2.0	0.5	2.6

and four more compensation capacitors used to compensate the frequency response of the V_{o-} terminal.

The dc transfer curves of opamp 3a were simulated using the circuit shown in Figure 8.9 with $R = 10K\Omega$. For $V_{i2} = 0.5$, 1.5, and $2.5V$, V_{i1} was swept from 0 to $3V$. The results are shown in Figure 8.12(a) where $V_{in} = V_{i1} - V_{i2}$ and $V_o = V_{o1} - V_{o2}$. Note that since $V_o = -V_{in}$ and since V_{in} can be either positive or negative, the input and the output range of the fully-differential opamps are twice as large as in the single-ended versions given in the previous section. The common mode voltage present at the input terminals of the opamp is shown in Figure 8.12(b). Note that a wide

Table 8.13: Deviations in the unity gain frequency of the two stage opamps with V_{CM} varied between 0.3 and 2.7V.

	opamp 2	opamp 2a	opamp 2b
$C_L(pF) = 5pF$, $f_{Umax}(MHz)$	1.88	1.80	1.81
$f_{Umin}(MHz)$	1.25	1.65	1.65
$f_{Uave}(MHz)$	1.49	1.71	1.73
$\%$ *deviation*	42	8.8	9.2
$C_L(pF) = 30pF$, $f_{Umax}(MHz)$	1.60	1.53	1.51
$f_{Umin}(MHz)$	1.02	1.33	1.37
$f_{Uave}(MHz)$	1.26	1.41	1.45
$\%$ *deviation*	46	14	9.7

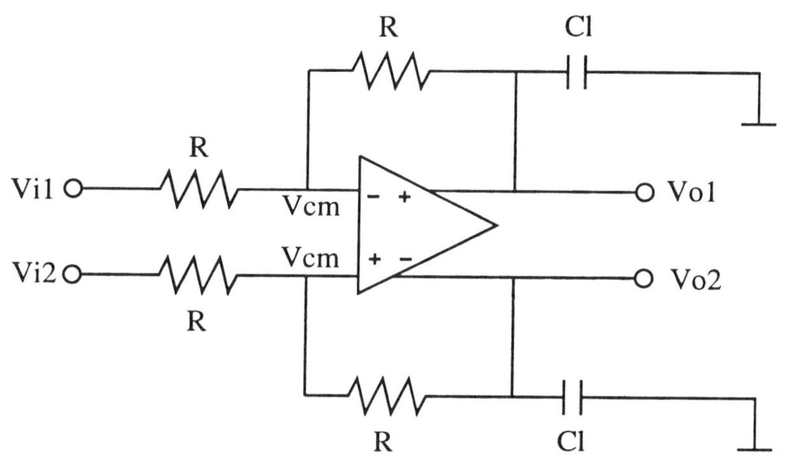

Figure 8.9: Inverting gain configuration used for the closed loop simulation of the fully-differential opamps.

range of V_{CM} is possible and that the constant-g_m input stage will still be useful for fully-differential opamps.

The small signal open loop frequency response of opamp 3a is shown in Figure 8.13. The dc components of V_{i1} and V_{i2} were set to $1.5V$ and the ac signal was applied at V_{i1}. The magnitude response at the V_{o1} and V_{o2} terminals are exactly the same except at very high frequencies, and there is an extra $6-dB$ of gain due to the differential output $V_{o1} - V_{o2}$. The phase responses at the two output terminals confirm the 180° phase shift as expected. More detailed simulation results are shown in Table 8.14 for

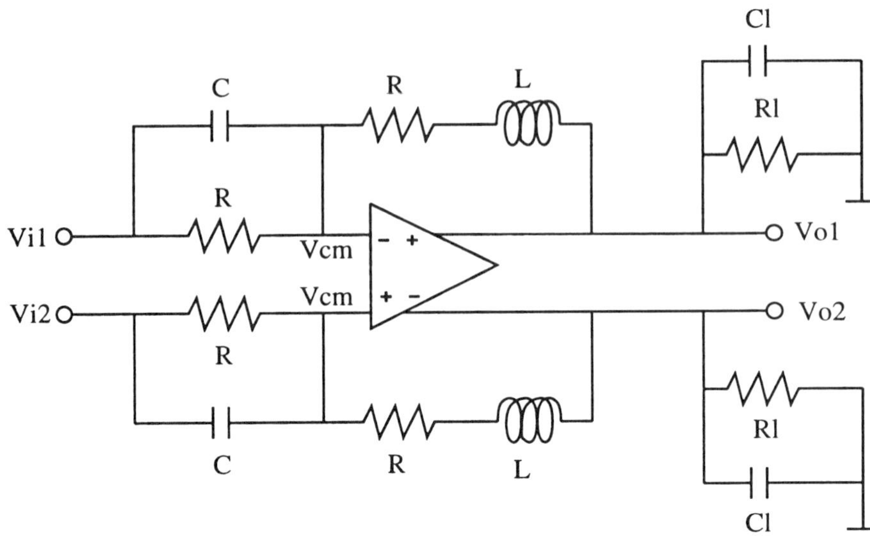

Figure 8.10: Circuit used for the open loop simulation of the fully-differential opamps.

different combinations of the input dc levels causing different V_{CM} at the input terminals of the opamp. The low frequency opamp gain, A_{DC}, and the unity gain frequency, f_U, were measured from the differential output whereas the phase margin and the gain margin were determined by studying the performance at both the positive and the negative output terminals to determine which one would cause instability. These combinations of the input voltages caused V_{CM} to be anywhere between $0.85V$ and $1.95V$, and A_{DC} and f_U are very flat except when the difference between the two inputs is large enough to push one of the output nodes close to the power supply rail and A_{DC} decreases slightly. Nonetheless, the constant-g_m input stage again helps to keep the unity gain frequency constant.

The common mode rejection ratio is measured as follows. Let A_{C1} and A_{C2} be the common mode gain from the input terminals of the opamp to the positive and the negative output terminals, respectively. Similarly, let A_{D1} and A_{D2} be the differential gains to the two output nodes. Then by applying the small signal, V_i, to both V_{i1} and V_{i2} in the inverting gain configuration of Figure 8.9, we get

$$\frac{V_{o1} - V_{o2}}{V_i} = \frac{A_{C1} - A_{C2}}{A_{D1} + A_{D2}} = CMRR \tag{8.2}$$

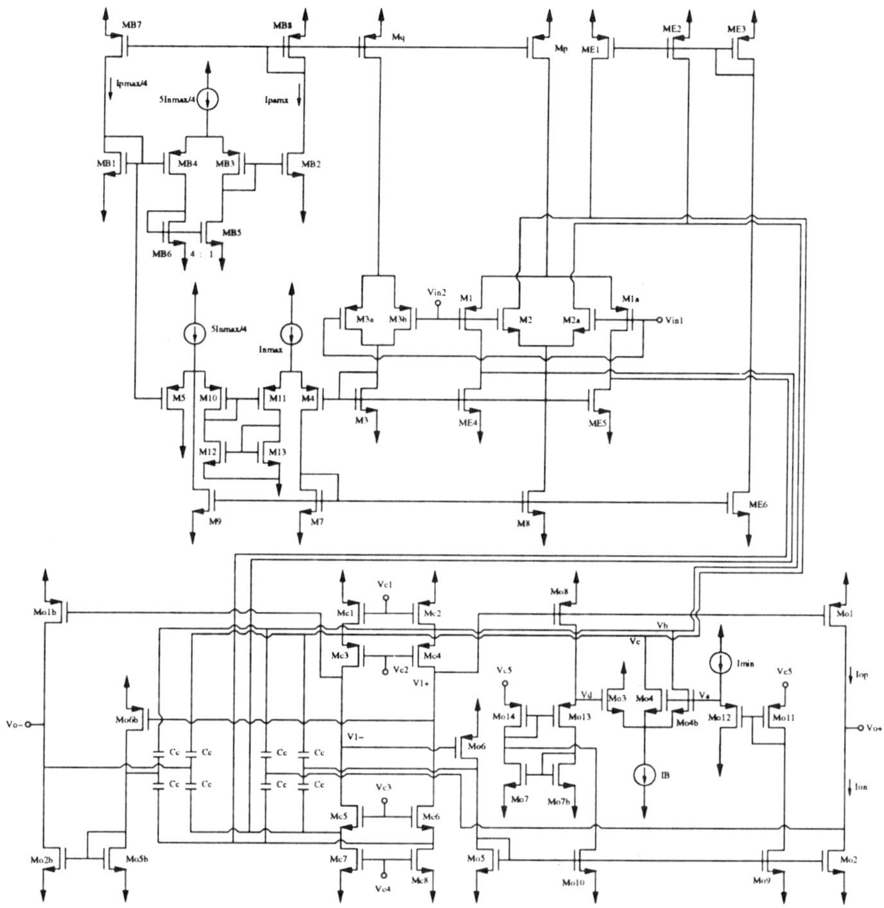

Figure 8.11: Opamp 3a: Fully-differential rail-to-rail two-stage opamp with the constant-g_m input stage 1.

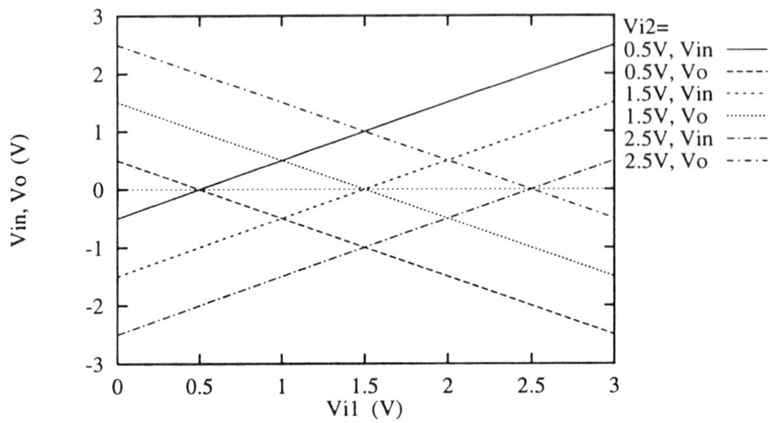

(a)

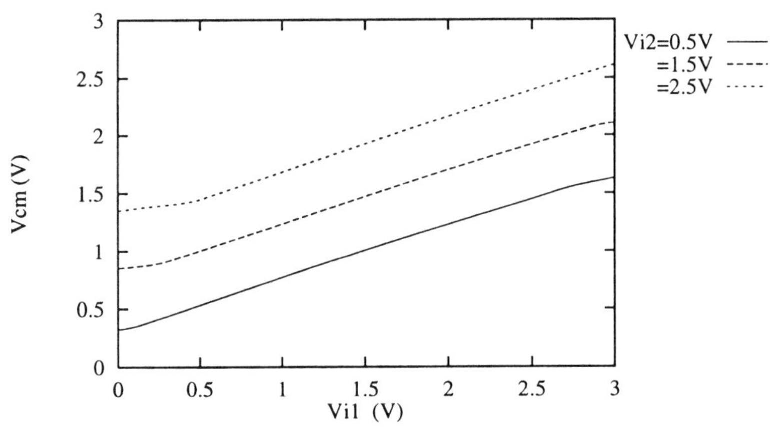

(b)

Figure 8.12: Simulation results of opamp 3a in an inverting gain configuration: (a) Vo-Vin1 characteristics, (b) Vcm-Vin1 characteristics.

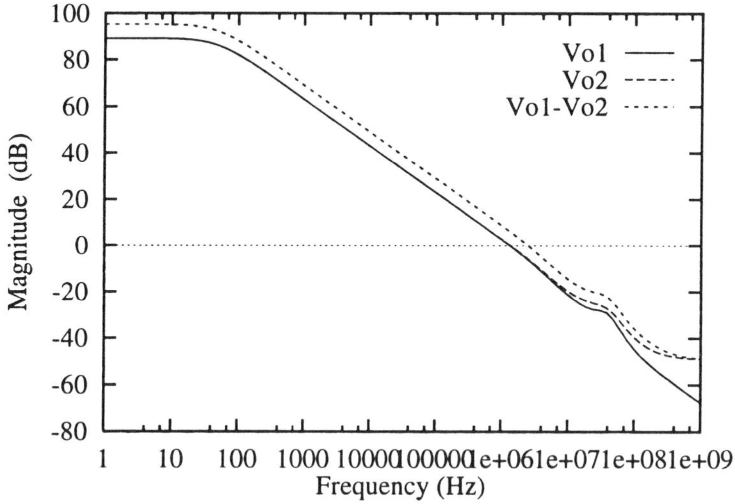

(a)

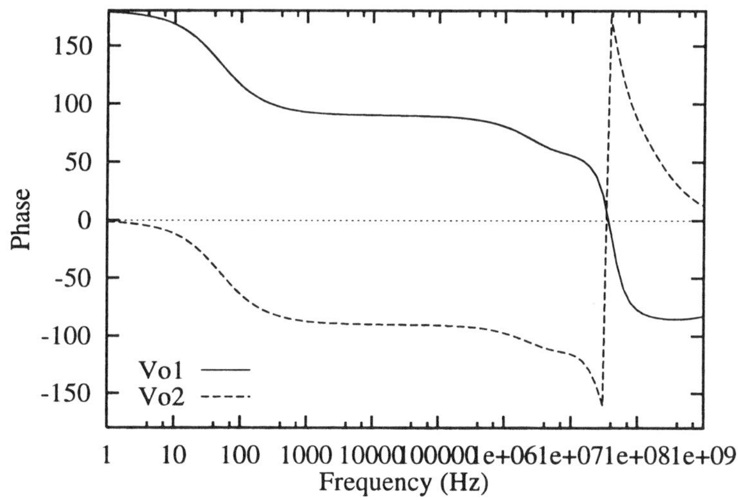

(b)

Figure 8.13: Open loop frequency response simulation of opamp 3a: (a) Magnitude response, (b) Phase response.

Table 8.14: Simulated frequency response of opamp 3a with $R_L = 100K\Omega$ and $C_L = 5pF$ at each output node.

V_{i1}	V_{i2}	V_{CM}	V_{O1}	V_{O2}	$A_{DC}(dB)$	$f_U(MHz)$	$\phi_M(^O)$	GM	f_{GM}
0.3	0.5	0.85	1.40	1.20	95	2.41	78	27	37
0.6	0.5	0.92	1.24	1.34	95	2.48	78	27	36
0.9	0.5	1.00	1.09	1.49	95	2.51	78	27	37
1.2	0.5	1.07	0.94	1.64	95	2.51	78	27	37
1.5	0.5	1.14	0.79	1.79	95	2.53	78	27	37
1.8	0.5	1.22	0.63	1.93	95	2.53	78	27	38
2.1	0.5	1.29	0.48	2.08	95	2.53	78	27	38
2.4	0.5	1.36	0.32	2.22	95	2.54	78	27	39
2.7	0.5	1.45	0.19	2.39	91	2.39	70	28	39
0.3	1.5	1.09	1.89	0.69	95	2.51	77	26	35
0.6	1.5	1.17	1.74	0.84	95	2.51	78	26	36
0.9	1.5	1.25	1.59	0.99	95	2.51	78	26	36
1.2	1.5	1.32	1.44	1.14	95	2.51	78	26	36
1.5	1.5	1.40	1.29	1.29	95	2.51	78	27	36
1.8	1.5	1.47	1.14	1.44	95	2.52	78	27	37
2.1	1.5	1.55	0.99	1.59	95	2.52	78	27	37
2.4	1.5	1.62	0.84	1.74	95	2.52	78	27	37
2.7	1.5	1.69	0.69	1.89	95	2.46	77	26	38
0.3	2.5	1.34	2.39	0.19	91	2.39	67	26	35
0.6	2.5	1.41	2.22	0.32	95	2.51	78	26	35
0.9	2.5	1.49	2.08	0.48	95	2.51	78	26	35
1.2	2.5	1.57	1.93	0.63	95	2.50	78	26	35
1.5	2.5	1.64	1.79	0.79	95	2.49	77	26	35
1.8	2.5	1.72	1.64	0.94	95	2.42	76	25	36
2.1	2.5	1.80	1.49	1.09	95	2.36	76	26	37
2.4	2.5	1.87	1.34	1.24	95	2.36	75	26	37
2.7	2.5	1.95	1.19	1.39	95	2.35	74	26	37

Table 8.15: Simulated common mode rejection ratio of opamp 3a with $C_L = 30pF$.

			$1/CMRR(dB)$ @(Hz)					
$V_{i1}(V)$	$V_{i2}(V)$	$V_{CM}(V)$	DC	$1K$	$10K$	$100K$	$1M$	$10M$
0.5	0.5	0.64	214	153	123	89	69	80
0.5	1.0	0.82	105	105	104	94	74	78
0.5	1.5	1.00	114	114	109	84	64	76
0.5	2.0	1.17	131	128	108	78	57	74
0.5	2.5	1.35	98	95	79	61	62	72
1.0	1.0	1.00	199	139	119	92	72	80
1.5	0.5	1.00	115	115	106	80	61	86
1.5	1.0	1.18	141	138	114	82	63	84
1.5	1.5	1.35	222	160	124	90	70	81
1.5	2.0	1.53	139	136	119	92	67	79
1.5	2.5	1.71	105	105	100	77	54	67
2.0	2.0	2.00	193	133	123	87	63	72
2.5	0.5	1.35	98	96	79	61	60	79
2.5	1.0	1.52	130	128	108	75	56	86
2.5	1.5	1.70	107	106	97	73	52	78
2.5	2.0	1.88	140	117	97	83	45	73
2.5	2.5	2.06	181	121	101	85	49	73

as long as the sum of the differential gains is much larger than one. Thus, if the common mode gain to the two terminals is exactly the same, then there should be an infinite rejection. This should take place when the two common source output stages are operating in exactly the same way, or when $V_{o1} = V_{o2}$; this of course happens when $V_{i1} = V_{i2}$(in the simulation, since the ac signals have no amplitude, $V_{i1} = V_{i2}$ means that their dc values are the same.) Simulation results are listed in Table 8.15 and we can clearly see that $CMRR$ is the best when $V_{i1} = V_{i2}$. Also, we can see that the same trend, present in the $CMRR$ of opamp 2a, can be seen here. That is $CMRR$ tends be be smaller when V_{CM} is close to $2.0V$. We can see this by checking $CMRR$ for the case $V_{i1} = V_{i2}$.

Simulation results of the power supply rejection ratio are shown in Table 8.16. For the same reason given for $CMRR$, $PSRR$ is the best when $V_{i1} = V_{i2}$.

The step response of opamp 3a was simulated by applying a step input at either one or both input terminals V_{i1} and V_{i2} in the configuration shown in Figure 8.9 with $R = 100K\Omega$. The first few rows in Table 8.17 shows the

Table 8.16: Simulated power supply rejection ratio of opamp 3a with $C_L = 30pF$.

			$1/PSRR(dB)$ from V_{DD} @(Hz)					
$V_{i1}(V)$	$V_{i2}(V)$	$V_{CM}(V)$	DC	$1K$	$10K$	$100K$	$1M$	$10M$
0.5	0.5	0.64	140	80	60	40	22	39
0.5	1.0	0.82	71	70	61	42	20	36
0.5	1.5	1.00	66	65	53	33	14	34
0.5	2.0	1.17	64	62	48	28	10	32
0.5	2.5	1.35	62	61	47	28	10	31
1.0	1.0	1.00	140	80	60	40	22	41
1.5	0.5	1.00	68	65	48	28	11	46
1.5	1.0	1.18	73	69	52	32	15	43
1.5	1.5	1.35	140	80	60	40	22	39
1.5	2.0	1.53	72	72	63	43	21	37
1.5	2.5	1.71	65	64	51	31	13	34
2.0	2.0	2.00	140	80	60	40	21	40
2.5	0.5	1.35	63	61	45	26	10	38
2.5	1.0	1.52	65	63	46	26	9	44
2.5	1.5	1.70	67	64	47	27	10	46
2.5	2.0	1.88	72	69	51	31	13	44
2.5	2.5	2.06	140	80	60	39	20	40
			$1/PSRR(dB)$ from V_{SS}					
0.5	0.5	0.64	135	75	55	36	17	33
0.5	1.0	0.82	68	68	56	37	16	33
0.5	1.5	1.00	64	63	48	28	10	33
0.5	2.0	1.17	63	60	44	24	6	33
0.5	2.5	1.35	62	59	42	23	5	33
1.0	1.0	1.00	136	76	56	36	18	33
1.5	0.5	1.00	68	62	44	24	7	33
1.5	1.0	1.18	71	66	48	28	10	33
1.5	1.5	1.35	136	76	56	36	17	33
1.5	2.0	1.53	71	70	58	38	17	33
1.5	2.5	1.71	63	62	47	26	8	33
2.0	2.0	2.00	136	76	56	36	17	33
2.5	0.5	1.35	64	59	41	21	5	29
2.5	1.0	1.52	65	60	41	21	4	33
2.5	1.5	1.70	67	61	43	23	6	33
2.5	2.0	1.88	70	65	47	26	9	33
2.5	2.5	2.06	136	76	56	35	16	33

results of the simulation with $0.4V$ step applied at V_{i1} while V_{i2} was held constant. The initial values of V_{i1} and V_{i2} used here cause the common mode voltage, V_{CM}, present at the input terminals of the opamp to be about $0.6V$, $1.0V$, and $2.0V$. Both $T_{S\uparrow}$ and $T_{S\downarrow}$ are about $0.5\mu sec$ for these inputs. For the other inputs listed, step sizes of $1.0V$ and $2.4V$ were applied at one or both input terminals, and it can be seen that the settling time increases only slightly when steps are applied at both inputs.

The distortion measurements shown in Table 8.18 were obtained by applying the sinusoid only at the V_{i1} terminal, except for one set of inputs that has a fully-balanced sinusoid. As expected, using a balanced input results in very low distortion as the even order harmonics are cancelled at the output nodes. The dc and ac components of V_{CM} produced by each input combination is also listed in the table. Since the outputs are taken differentially, the large inputs, which resulted in large THD for opamp 2a, result in much less THD in opamp 3a.

8.2.2 Opamp 3b: Fully-Differential Rail-to-Rail Folded - Cascode Opamp With Constant-g_m Input Stage 2

Figure 8.14 shows the fully-differential version of opamp 2b. Again, the only difference between opamps 2b and 3b is the addition of four transistors M_{o1b}, M_{o2b}, M_{o5b}, and M_{o6b}, and four extra compensation capacitors.

Table 8.19 shows the results of the open loop frequency response; we see that as V_{CM} changes between about $0.9V$ and $2.0V$, A_{DC} and f_U stay fairly constant thanks to the constant-g_m input stage 2. Simulated quiescent power dissipation of the opamps are $1.2mW$ and $1.3mW$ for opamp 3a and opamp 3b, respectively.

The common mode and the power supply rejection ratios are shown in Table 8.20 and Table 8.21, respectively. As was the case with opamp 3a, both $CMRR$ and the $PSRR$ are the best when $V_{i1} = V_{i2}$; also, they follow the behavior seen in its single-ended output version.

Table 8.22 lists the settling time of opamp 3b, and shows that both $T_{S\uparrow}$ and the $T_{S\downarrow}$ are fairly constant for $0.4V$ input step regardless of the value of V_{CM} present at the opamp input terminals.

Table 8.23 lists the total harmonic distortion of opamp 3b for various input voltage combinations.

Table 8.17: Simulation results of 2% settling time of opamp 3a with $C_L = 5pF$.

$V_{i1\,STEP}$	$V_{i1\,INIT}$	$V_{i2\,STEP}$	$V_{i2\,INIT}$	$T_{S\uparrow}(\mu sec)$	$T_{S\downarrow}(\mu sec)$	$P.O_{\uparrow}$	$P.O_{\downarrow}(\%)$
0.4	0.3	0.0	0.5	0.49	0.50	5.0	5.0
0.4	1.3	0.0	1.5	0.48	0.48	7.5	7.5
0.4	2.3	0.0	2.5	0.50	0.52	7.5	10
1.0	1.0	0.0	1.5	0.50	0.49	6.0	6.0
1.0	1.0	-1.0	2.0	0.53	0.54	4.0	3.5
2.4	0.3	0.0	1.5	0.61	0.50	2.9	3.3
2.4	0.3	-2.4	2.7	0.64	0.65	1.0	0.8

Table 8.18: Simulated total harmonic distortion of opamp 3a with $C_L = 5pF$.

$f_O(Hz)$	$V_{i1\,DC}$	$V_{i1\,AC}$	$V_{i2\,DC}$	$V_{i2\,AC}$	$V_{CM\,DC}$	$V_{CM\,AC}$	Thd(%)
10K	1.0	0.5	1.0	0.0	1.00	0.18	0.010
10K	1.5	0.5	1.5	0.0	1.35	0.18	0.005
10K	2.0	0.5	2.0	0.0	1.71	0.18	0.021
10K	1.5	1.0	1.5	0.0	1.35	0.35	0.003
10K	1.5	1.0	1.5	-1.0	1.35	0.01	0.001
10K	1.5	1.5	1.5	0.0	1.35	0.52	0.014
100K	1.0	0.5	1.0	0.0	1.00	0.17	0.086
100K	1.5	0.5	1.5	0.0	1.35	0.17	0.009
100K	2.0	0.5	2.0	0.0	1.71	0.17	0.021
100K	1.5	1.0	1.5	0.0	1.35	0.34	0.024
100K	1.5	1.5	1.5	0.0	1.35	0.51	0.11
200K	1.0	0.5	1.0	0.0	1.00	0.15	0.13
200K	1.5	0.5	1.5	0.0	1.35	0.16	0.040
200K	2.0	0.5	2.0	0.0	1.71	0.16	0.44
200K	1.5	1.0	1.5	0.0	1.35	0.31	0.081
200K	1.5	1.5	1.5	0.0	1.35	0.47	0.20

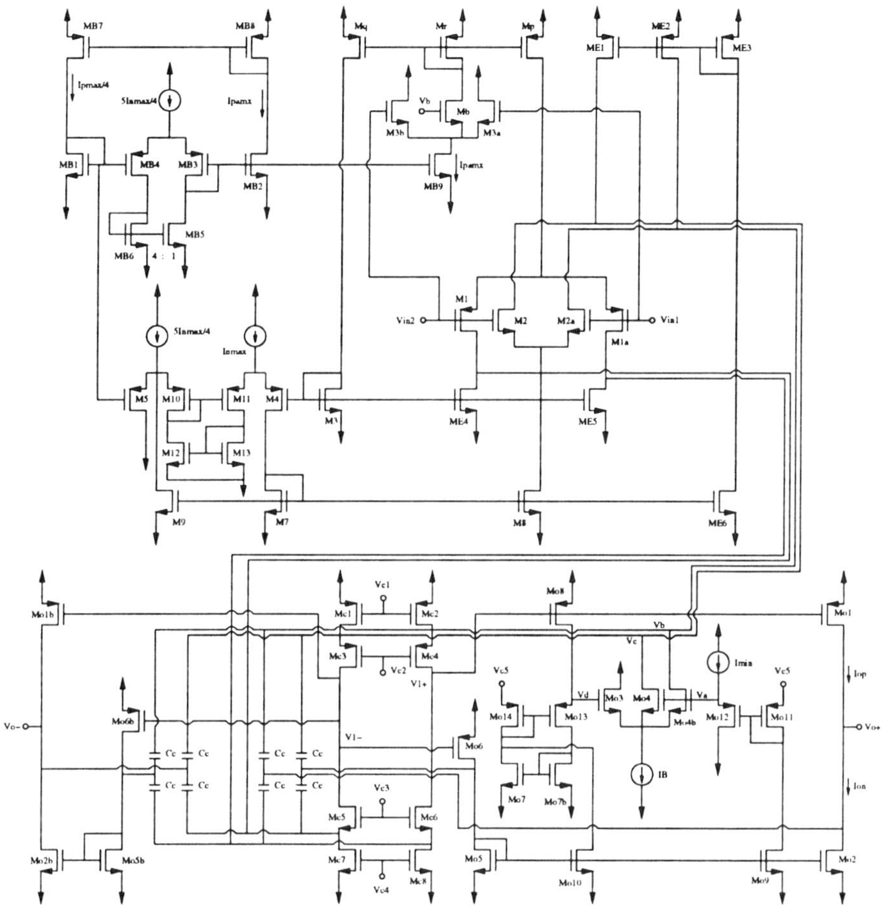

Figure 8.14: Opamp 3b: Fully-differential rail-to-rail two-stage opamp with the constant-g_m input stage 2.

Table 8.19: Simulated frequency response of opamp 3b with $R_L = 100K\Omega$ and $C_L = 5pF$ at each output node.

V_{i1}	V_{i2}	V_{CM}	V_{O1}	V_{O2}	$A_{DC}(dB)$	$f_U(MHz)$	$\phi_M(^O)$	GM	f_{GM}
0.3	0.5	0.85	1.39	1.19	95	2.50	79	27	38
0.6	0.5	0.92	1.24	1.34	95	2.56	79	27	38
0.9	0.5	1.00	1.09	1.49	95	2.59	79	27	39
1.2	0.5	1.07	0.94	1.64	96	2.59	79	27	39
1.5	0.5	1.14	0.79	1.79	96	2.60	79	27	40
1.8	0.5	1.22	0.63	1.93	96	2.59	79	27	40
2.1	0.5	1.29	0.48	2.08	95	2.52	79	28	39
2.4	0.5	1.36	0.32	2.22	95	2.43	79	28	39
2.7	0.5	1.45	0.20	2.40	90	2.22	70	29	39
0.3	1.5	1.09	1.88	0.68	96	2.59	79	26	38
0.6	1.5	1.17	1.74	0.84	96	2.59	79	26	38
0.9	1.5	1.24	1.59	0.99	94	2.55	79	27	38
1.2	1.5	1.32	1.44	1.14	95	2.46	79	27	38
1.5	1.5	1.40	1.29	1.29	95	2.38	79	27	37
1.8	1.5	1.47	1.14	1.44	95	2.35	77	28	37
2.1	1.5	1.54	0.99	1.59	95	2.36	74	28	36
2.4	1.5	1.62	0.84	1.74	94	2.36	74	28	35
2.7	1.5	1.69	0.68	1.88	94	2.34	74	26	35
0.3	2.5	1.35	2.40	0.20	90	2.26	65	27	37
0.6	2.5	1.41	2.22	0.32	95	2.36	78	27	36
0.9	2.5	1.49	2.08	0.48	95	2.33	76	27	36
1.2	2.5	1.57	1.93	0.63	94	2.33	74	27	35
1.5	2.5	1.64	1.79	0.79	94	2.33	73	26	34
1.8	2.5	1.72	1.64	0.94	94	2.33	74	28	34
2.1	2.5	1.80	1.49	1.09	94	2.41	77	28	35
2.4	2.5	1.87	1.34	1.24	94	2.48	78	28	35
2.7	2.5	1.95	1.19	1.39	95	2.46	78	28	35

Table 8.20: Simulated common mode rejection ratio of opamp 3b with $C_L = 30pF$.

			$1/CMRR(dB)$	$@(Hz)$				
$V_{i1}(V)$	$V_{i2}(V)$	$V_{CM}(V)$	DC	$1K$	$10K$	$100K$	$1M$	$10M$
0.5	0.5	0.67	188	128	108	85	67	81
0.5	1.0	0.84	106	105	95	76	63	87
0.5	1.5	1.02	121	117	99	77	59	88
0.5	2.0	1.18	116	116	107	77	56	91
0.5	2.5	1.36	99	97	81	61	46	74
1.0	1.0	1.02	188	128	108	86	67	81
1.5	0.5	1.01	119	119	109	85	63	76
1.5	1.0	1.19	125	125	119	98	69	84
1.5	1.5	1.36	182	122	102	81	67	82
1.5	2.0	1.53	114	110	93	72	52	77
1.5	2.5	1.70	95	95	94	87	48	80
2.0	2.0	1.97	189	129	109	93	58	74
2.5	0.5	1.36	99	96	80	60	46	71
2.5	1.0	1.52	103	102	90	69	49	67
2.5	1.5	1.70	95	95	95	94	52	69
2.5	2.0	1.87	102	102	102	105	68	76
2.5	2.5	2.04	203	143	121	85	64	76

Table 8.21: Simulated power supply rejection ratio of opamp 3b with $C_L = 30pF$.

			$1/PSRR(dB)$ from V_{DD} @(Hz)					
$V_{i1}(V)$	$V_{i2}(V)$	$V_{CM}(V)$	DC	$1K$	$10K$	$100K$	$1M$	$10M$
0.5	0.5	0.67	140	80	60	41	27	35
0.5	1.0	0.84	77	72	54	34	22	39
0.5	1.5	1.02	72	69	52	32	19	43
0.5	2.0	1.18	69	67	50	30	17	40
0.5	2.5	1.36	67	65	49	29	13	37
1.0	1.0	1.02	141	81	61	41	27	35
1.5	0.5	1.01	72	71	60	41	24	30
1.5	1.0	1.19	77	77	76	60	29	32
1.5	1.5	1.36	141	81	61	41	26	36
1.5	2.0	1.53	78	72	53	33	20	40
1.5	2.5	1.70	68	66	49	29	15	44
2.0	2.0	1.97	140	80	60	40	25	36
2.5	0.5	1.36	66	65	53	33	15	27
2.5	1.0	1.52	65	65	52	32	16	29
2.5	1.5	1.70	68	67	55	35	19	30
2.5	2.0	1.87	74	74	73	59	28	33
2.5	2.5	2.04	141	81	61	40	26	35
			$1/PSRR(dB)$ from V_{SS}					
0.5	0.5	0.67	134	74	54	34	21	26
0.5	1.0	0.84	83	67	48	28	15	26
0.5	1.5	1.02	75	65	45	26	13	27
0.5	2.0	1.18	71	63	44	24	11	27
0.5	2.5	1.36	72	62	43	23	6	25
1.0	1.0	1.02	135	75	55	35	21	26
1.5	0.5	1.01	74	71	54	34	19	25
1.5	1.0	1.19	75	75	75	56	24	26
1.5	1.5	1.36	135	78	55	35	20	27
1.5	2.0	1.53	92	68	48	28	15	27
1.5	2.5	1.70	73	65	46	26	12	27
2.0	2.0	1.97	137	77	57	37	22	27
2.5	0.5	1.36	69	65	46	26	9	27
2.5	1.0	1.52	67	63	46	26	12	26
2.5	1.5	1.70	71	69	52	32	17	27
2.5	2.0	1.87	79	79	76	57	29	28
2.5	2.5	2.04	140	80	60	40	25	28

Table 8.22: Simulation results of 2% settling time of opamp 3b with $C_L = 5pF$.

$V_{i1\ STEP}$	$V_{i1\ INIT}$	$V_{i2\ STEP}$	$V_{i2\ INIT}$	$T_{S\uparrow}(\mu sec)$	$T_{S\downarrow}(\mu sec)$	$P.O_{\uparrow}$	$P.O_{\downarrow}(\%)$
0.4	0.3	0.0	0.5	0.44	0.43	3.5	2.5
0.4	1.3	0.0	1.5	0.46	0.47	4.5	3.8
0.4	2.3	0.0	2.5	0.42	0.41	3.0	3.0
1.0	1.0	0.0	1.5	0.52	0.47	4.0	3.0
1.0	1.0	-1.0	2.0	0.51	0.51	1.8	1.5
2.4	0.3	0.0	1.5	0.59	0.55	2.1	1.0
2.4	0.3	-2.4	2.7	0.86	0.85	1.0	0.8

Table 8.23: Simulated total harmonic distortion of opamp 3b with $C_L = 5pF$.

$f_O(Hz)$	$V_{i1\ DC}$	$V_{i1\ AC}$	$V_{i2\ DC}$	$V_{i2\ AC}$	$V_{CM\ DC}$	$V_{CM\ AC}$	$THD(\%)$
10K	1.0	0.5	1.0	0.0	1.02	0.17	0.005
10K	1.5	0.5	1.5	0.0	1.36	0.17	0.025
10K	2.0	0.5	2.0	0.0	1.70	0.17	0.019
10K	1.5	1.0	1.5	0.0	1.36	0.35	0.031
10K	1.5	1.0	1.5	-1.0	1.36	0.01	0.002
10K	1.5	1.5	1.5	0.0	1.35	0.51	0.031
100K	1.0	0.5	1.0	0.0	1.00	0.16	0.042
100K	1.5	0.5	1.5	0.0	1.36	0.16	0.26
100K	2.0	0.5	2.0	0.0	1.70	0.17	0.20
100K	1.5	1.0	1.5	0.0	1.36	0.33	0.32
100K	1.5	1.5	1.5	0.0	1.35	0.50	0.33
200K	1.0	0.5	1.0	0.0	1.02	0.15	0.063
200K	1.5	0.5	1.5	0.0	1.36	0.15	0.53
200K	2.0	0.5	2.0	0.0	1.70	0.16	0.41
200K	1.5	1.0	1.5	0.0	1.36	0.31	0.72
200K	1.5	1.5	1.5	0.0	1.36	0.45	0.80

Chapter 9

Silicon Implementations

This chapter contains results of measurements performed on fabricated low-voltage operational amplifiers. Eight opamps were implemented onto two MOSIS Tiny chips. Each chip was fabricated in a $2\mu m$ p-well process on the same run; the process parameters from this run are given in Appendix A. All the opamps were tested at dc to determine the input and output voltage range. The dc offsets were measured as a function of the common mode input voltage. The input stages of the single-stage opamps were used to measure the input stage current and evaluate the performance of the input transconductance. All dc measurements were performed using an HP4145 semiconductor parameter analyzer with 3-V power supply, except for the fully-differential opamps which were test with 3.3-V power supply.

The frequency response of the opamps discussed in this chapter is characterized by measuring the low frequency open loop gain, the unity gain frequency, and the phase margin. We have noticed that using 3-V power supply caused the open loop gain to be 20 to 30dB less than the expected values. We were able to regain the low frequency gain by increasing the power supply to above 3.15V, thus, we used 3.3V for these measurements. Although we were able to obtain the expected results, some care had to be taken in order to have the constant-g_m input stage operating properly. First, we noticed that the constant-g_m input stage which worked when measured with HP4145 did not have the constant-g_m characteristics when it was powered by the HP6237B power supplies. Second, constant-g_m input stages that worked on HP4145 when sweeping the common mode input voltage initially from $V_{SS} = 0V$ to $V_{DD} = 3V$ did not work when the input voltage was swept from somewhere near the mid-rail to V_{DD}. However, we were able to find ranges of power supply voltages, other than zero to 3V, on the HP4145 that resulted in the proper operation of the input stage, that is, we kept $V_{DD} - V_{SS} = 3.3V$, but started the sweep from a value other than $V_{SS} = 0$. More details are discussed in the following sections. The problem is believed to be associated with the transient response at the power up, and simulation results which support the observation are provided along with a simple solution to the problem in the last section of this chapter.

The 2% settling time and the harmonic distortions of the opamps were measured with the same biasing condition used for the frequency response measurements, *i. e.*, the circuits were powered up with the HP4145. The only experimental results obtained on the fully-differential opamps were of the dc measurements because of the parasitic poles that exist at the input terminals of the opamp when used in the resistive circuit as shown in Figure 8.9. The methods used for the measurements of the above mentioned opamp characteristics are provided in Appendix C.

9.1 Chip Organization

The photomicrographs of two Tiny chips containing 3-V operational amplifiers are shown in Figure 9.1. Chip 1, as shown in Figure 9.1(a) and in Figure 9.2, contains three single-stage opamps, opamp 1, opamp 1a, and opamp 1b, and one two-stage opamp, opamp 2. It also contains p- and n-channel differential pairs where all of the terminals are accessible. All single-stage opamps have two extra transistors connected to them, an NMOS whose gate and source, respectively, are connected to the gate and the source, respectively, of the NMOS current source providing bias current to the n-channel input differential pair, and a PMOS that is connected similarly to the PMOS current source. The purpose of these transistors is, of course, to test the constant-g_m input stages; this is done together with the separate differential pairs placed on chip 1. Constant bias currents I_{nmax} and I_{C1}, and constant bias voltages V_{C2} and V_{C3} are all shared between the four opamps. Opamp 1b has an extra pin for V_b which is used as constant bias voltage for input stage 2. Opamp 2 also has an extra pin for V_{C5} which is used to bias the class AB output stage. The current entering V_{C5} together with the output current are used later to determine the current flow of the n- and p-channel transistors in the common source output stage.

Chip 2, as shown in Figure 9.1(b) and in Figure 9.3, contains two-stage opamps with constant-g_m input stages. Two of them are single-ened output opamps, opamp 2a and opamp 2b, and the other two are fully-differential opamps, opamp 3a and opamp 3b. The chip also contains single transistors, in case the need for device characterization arises. Similarly to the chip 1, the constant biases I_{nmax}, I_{min}, I_{C1}, V_{C2}, and V_{C3} are all shared among

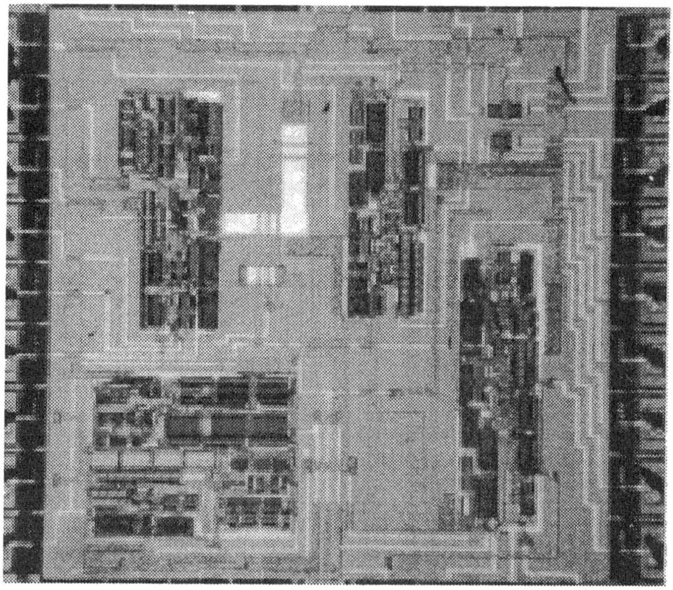

(a)

(b)

Figure 9.1: Photomicrographs of the fabricated chips. (a) Chip 1. (b) Chip 2.

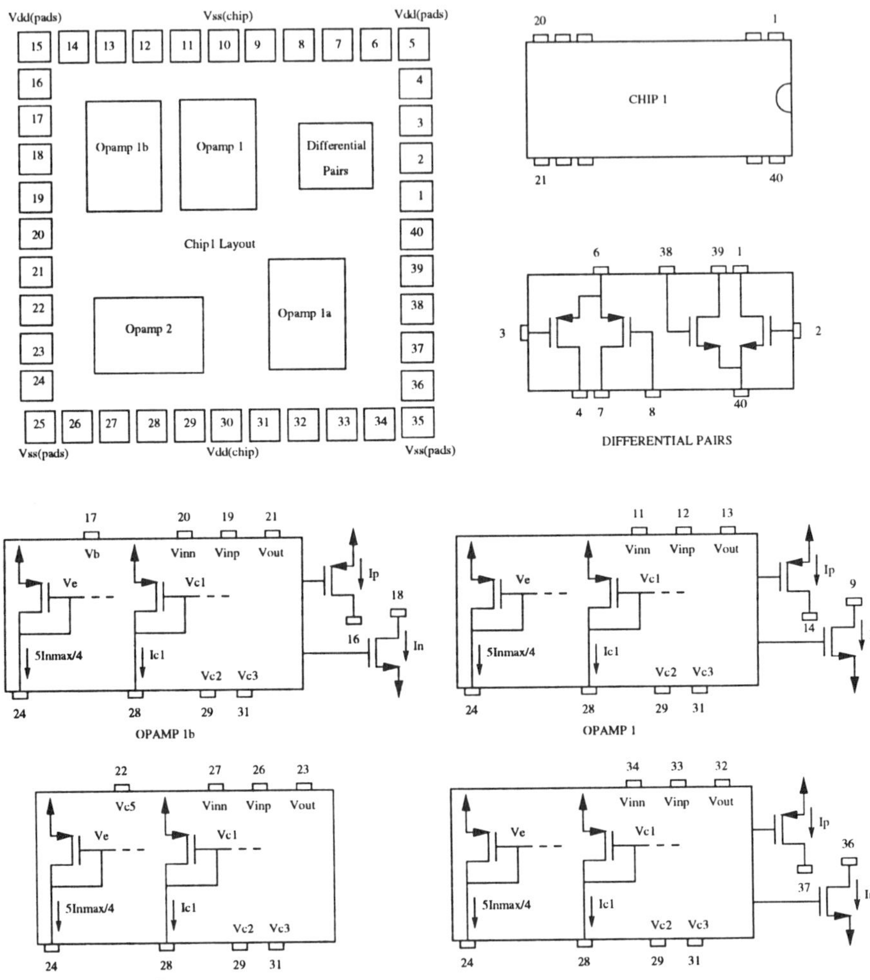

Figure 9.2: Organization of chip 1.

Table 9.1: Area occupied by each opamp.

Name	Input Stage	Architecture	Area $\mu m \times \mu m$
Opamp 1	rail-to-rail	single stage, single-ended	750 × 250
Opamp 1a	input stage 1	single stage, single-ended	870 × 250
Opamp 1b	input stage 2	single stage, single-ended	760 × 340
Opamp 2	rail-to-rail	two stage, single-ended	650 × 510
Opamp 2a	input stage 1	two stage, single-ended	705 × 560
Opamp 2b	input stage 2	two stage, single-ended	670 × 590
Opamp 3a	input stage 1	two stage, fully differential	705 × 630
Opamp 3b	input stage 2	two stage, fully differential	745 × 590

the four opamps. Opamps 2b and 3b, which contain input stage 2, each has its own V_b bias voltage, and each of the four opamps has its own V_{C5} bias voltage.

The size of each opamp is added to Table 6.1 and is shown in Table 9.1. Opamps 1 and 2, which do not possess a constant-g_m, were biased by I_{nmax} and I_{pmax} generated by the bias circuit of Figure 5.6; this makes these opamps slightly larger than if they were biased simply by I_{ref} as shown in Figure 7.1. Note that the presence of the output stage doubles the size of an opamp while the difference in the size of the opamps with single-ended and fully-differential output stages is very small.

9.2 Input Stages

As mentioned in the previous section, the input stage characteristics will be extracted from the input stages of the single-stage opamps placed on chip 1. The results given in this section were obtained with input stages biased with $I_{nmax} = 18\mu A$ and powered with $V_{DD} = 3V$ and $V_{SS} = 0V$. Transistor sizes of the differential pairs are $W/L = 75\mu m/8\mu m$ and $30\mu m/8\mu m$ for the p- and n-channel transistors, respectively. The ratio of 2.5 for the size of these transistors was chosen because it is the ratio of the n-type to the p-type transconductance parameters given in the parameter set used for the design of the opamps. Note that the ratio was the same on the parameter set provided by MOSIS for the process which our chips were fabricated with. Thus, the K_p to K_n ratio is expected to be close to one. In order to determine the true value of the $K_p - K_n$ ratio, the drain currents of n- and p-channel transistors were measured as a function of the gate to source voltage. This is shown in Figure 9.4(a), and the square roots of the currents are shown in Figure 9.4(b); since the transistors are in the

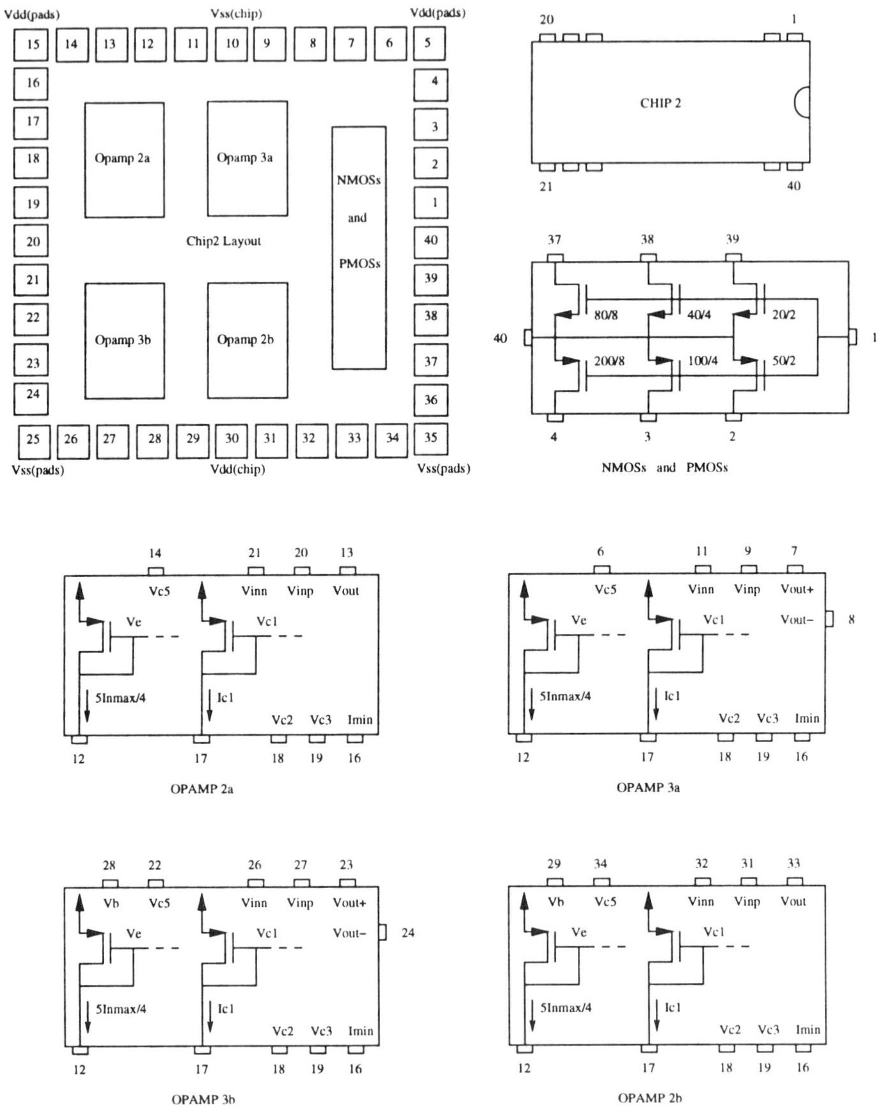

Figure 9.3: Organization of chip 2.

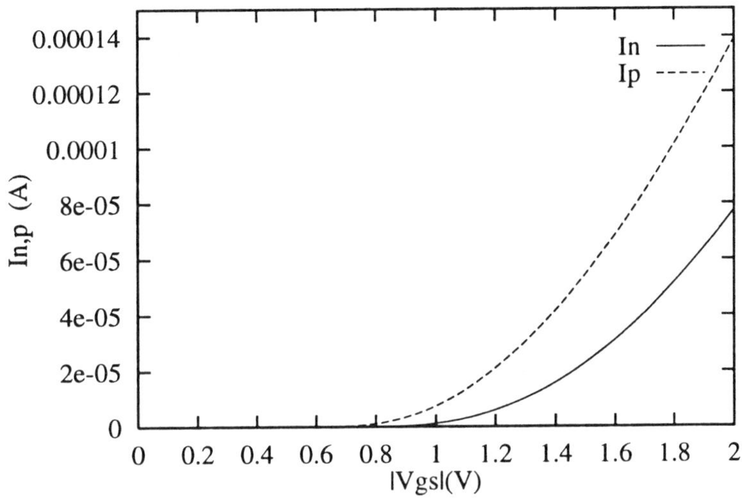

(a)

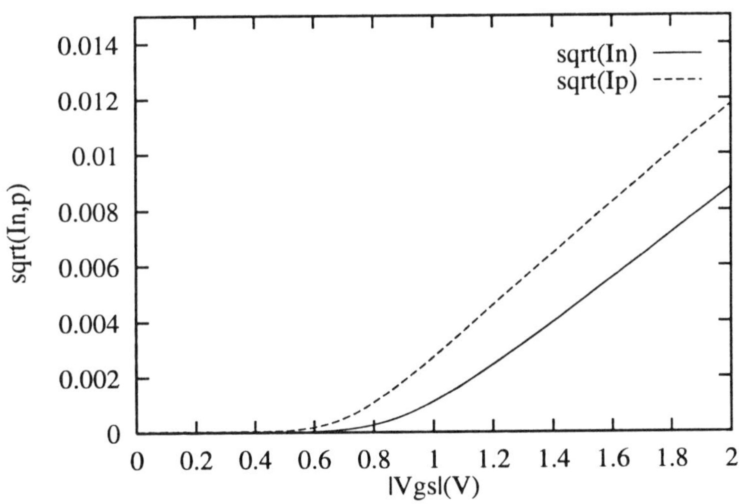

(b)

Figure 9.4: Drain currents of the transistors used in the differential pairs. (a) I_n and I_p. (b) Square roots of I_n and I_p.

saturation region, K_n and K_p can be determined from the slope of these curves. We found that $K_n = 52\mu A/V^2$ and $K_p = 87\mu A/V^2$, and $K_p - K_n$ ratio is 1.67 instead of the expected value of 1.

The uncertainty of the $K_p - K_n$ ratio is precisely the reason that we are developing the constant-g_m input stage which does not require the matching of K_p and K_n, and in the following we provide the experimental results of the input stages fabricated in $2\mu m$ MOSIS process.

9.2.1 Input Stage Without the Constant-g_m Bias Circuit

The input stage used in opamp 1 does not have the constant-g_m characteristics; however, the bias currents of the differential pairs are provided by the circuit that was given in Figure 5.6 and satisfy the condition given by (5.10) such that $g_{mT} = g_{mp}$ when V_{CM} is near V_{SS} would equal to $g_{mT} = g_{mn}$ when V_{CM} is near V_{DD}. Figure 9.5(a) shows the differential pair currents; the currents plotted are of one transistor of the differential pair. Note that I_p is significantly less than I_n which reveals that K_p is larger than K_n and that is why I_p must be smaller in order to satisfy (5.10). Figure 9.5(b) shows the measured g_m of the input stage which does not possess the constant-g_m characteristics. We see that g_{mT} varies by a factor of two as expected; moreover, because of the bias circuit of Figure 5.6, g_{mT} is the same for V_{CM} near the two rails despite the fact that K_p/K_n is not close to one.

9.2.2 Constant-g_m Input Stage 1

Figure 9.6(a) shows the differential pair currents of the constant-g_m input stage 1. Note that these currents are the sum of currents in both transistors for each differential pair, and hence are twice as much as the currents shown in Figure 9.5(a). Figure 9.6(b) shows the measured g_m of the differential pairs, and the curves closely resemble the simulation results. g_{mT} is constant within about 10% for the entire common mode range.

9.2.3 Constant-g_m Input Stage 2

Constant-g_m input stage 2 used as the input stage for opamp 1b, was tested in exactly the same way that constant-g_m input stage 1 was tested. However, this circuit has an external bias voltage, V_b, in the current monitoring part of the circuit and we have measured the currents and the transconductances of the input stage for three different values of V_b. They are shown in

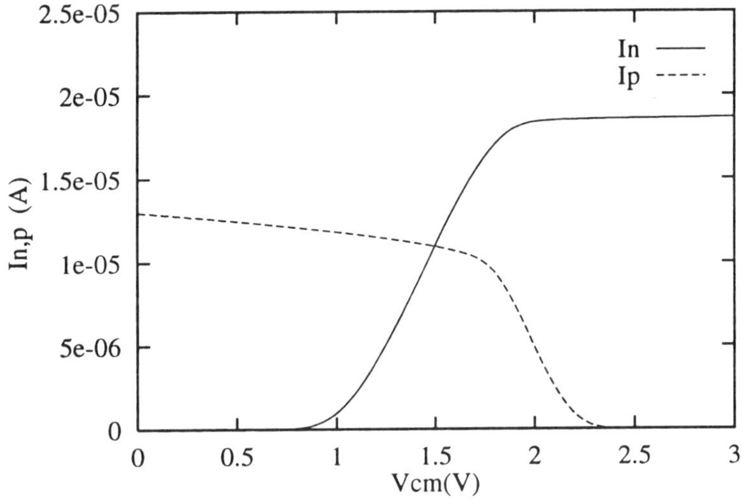

(a)

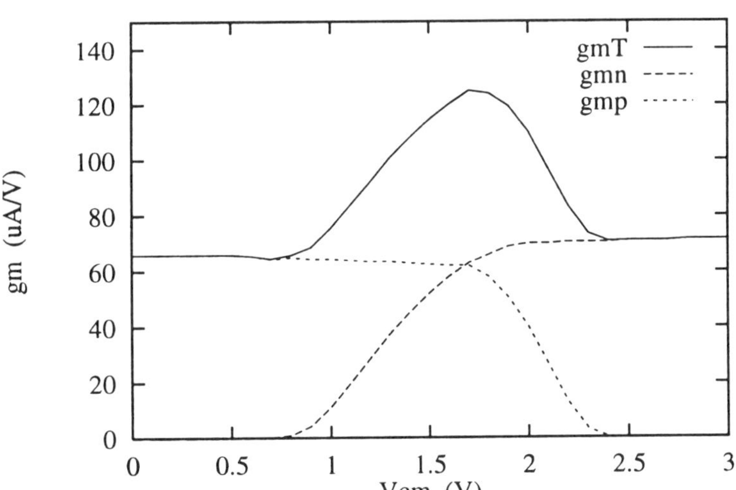

(b)

Figure 9.5: Measurements taken on the input stage of opamp 1 . (a) Differential pair currents. (b) Differential pair transconductance.

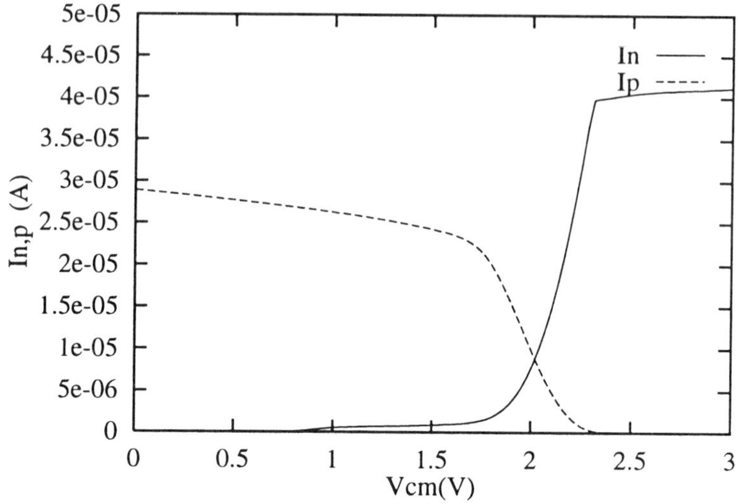

(a)

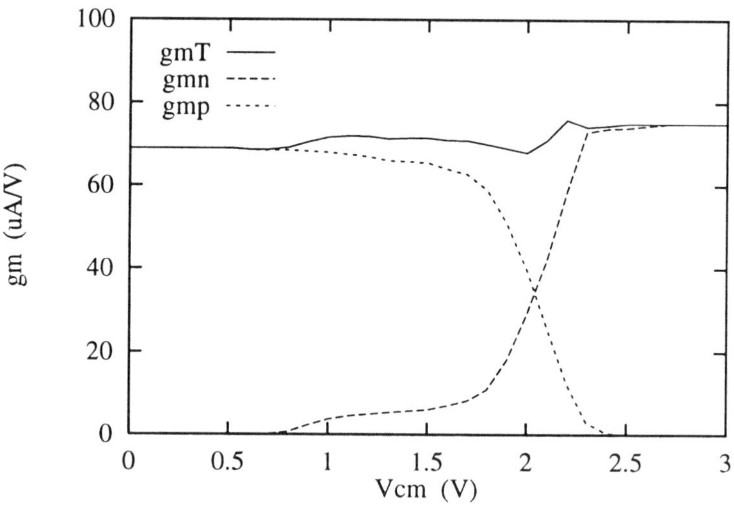

(b)

Figure 9.6: Measurements taken on the input stage of opamp 1a . (a) Differential pair currents. (b) Differential pair transconductance.

Figure 9.7 and we can clearly see the dependence of the input stage characteristics on the value of V_b. Again, the results are similar to the simulation results even to the point that I_n in this circuit reaches I_{nmax} less sharply than in the constant-g_m input stage 1.

Even though the results provided here are obtained using $V_{DD} - V_{SS} = 3V$, we observed that both constant-g_m input stages operate properly even with $V_{DD} - V_{SS} = 2.5V$. This observation should come as no surprise since, as discussed earlier in Section 4.1, the minimum supply voltage required is two gate-to-source voltage drops(in strong inversion) plus V_{DSsat}. This makes the new input stages yet more attractive, especially since the use of a $2.7 - V$ power supply is to follow that of a $3 - V$ supply in the future.

9.3 Single-Stage Operational Amplifiers

9.3.1 dc Measurements

For $V_{DD} = 3.0V$ and $V_{SS} = 0V$, all single stage opamps were biased with $V_{C2} = V_{C3} = 1.5V$, $I_{nmax} = 18\mu A$, and $I_{C1} = 55\mu A$. The input-output characteristics of opamp 1 in a unity gain configuration are shown in Figure 9.8(a); the curves for opamp 1a and opamp 1b are identical to that of opamp 1. The output range of the single-ended opamps is from $0.5V$ to $2.3V$ as expected. The offset voltages of the opamps are shown in Figure 9.8(b). While the offest voltages of opamp 1 and opamp 1a change from about $-10mV$ at $V_{in} = .5V$ to about $0mV$ at $V_{in} = 2.3V$, the offset voltage of opamp 1b is about $10mV$ shifted toward V_{SS} in that range.

9.3.2 Frequency Response

As mentioned at the beginning of this chapter, the low frequency gain was significantly less than expected for $V_{DD} - V_{SS} = 3.0V$, and we had to increase the supply voltage to more than $3.15V$ to restore the higher gain as expected. This increase in supply voltage placed all the transistors in the cascode stage in the saturation region. Thus, all the measurement results provided in the rest of this section were obtained using $V_{DD} - V_{SS} = 3.3V$. Furthermore, the bias for the cascode stage were changed to $V_{C2} = 1.75V$, $V_{C3} = 1.65V$, and $I_{C1} = 37\mu A$. Also, we noticed that certain values of V_{DD} and V_{SS} caused I_n of the constant-g_m input stage to be held at I_{nmax}. That is, the input stages of opamp 1a and opamp 1b were behaving exactly like that of the opamp 1. Nonetheless, we were able to find a good set of V_{DD} and V_{SS} which provided the measurements we expected. Note that once a good set was found for a given circuit, the same set was used for all of the

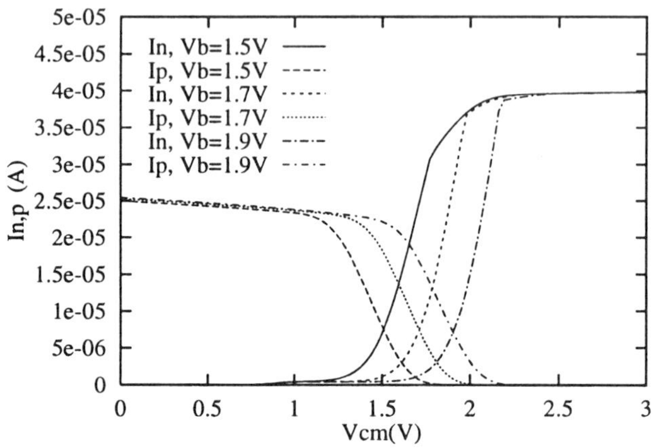

(a)

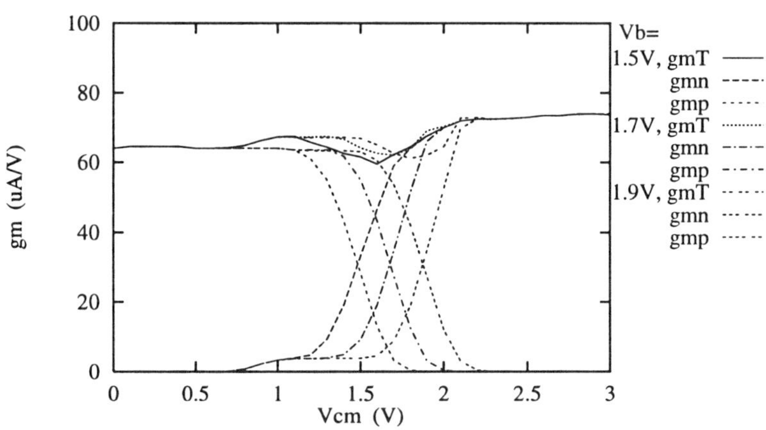

(b)

Figure 9.7: Measurements taken on the input stage of opamp 1b . (a) Differential pair currents. (b) Differential pair transconductance.

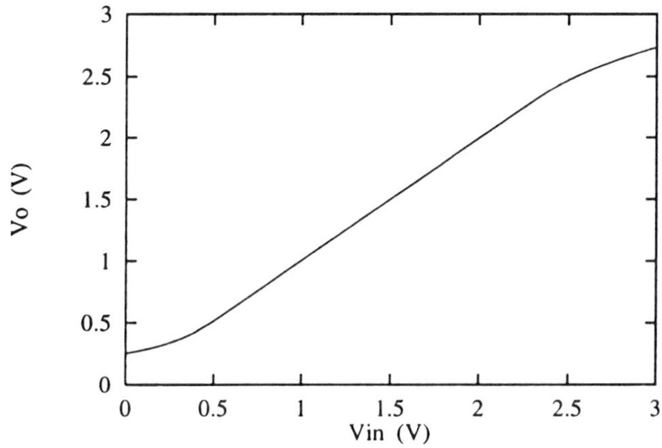

(a)

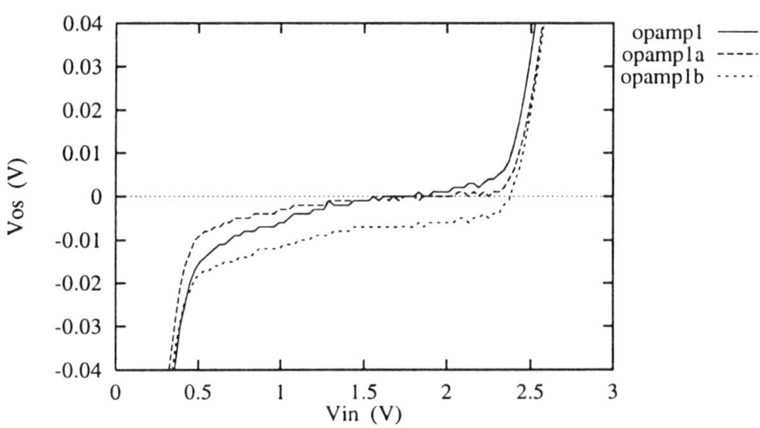

(b)

Figure 9.8: Experimental results of the single-stage opamps in a unity gain configuration. (a) Vin-Vo characteristics. (b) Offset voltages.

measurements. All the dc voltages given in the following discussions are with respect to V_{SS}.

Table 9.2 shows the low frequency gain, A_{DC}, the unity gain frequency, f_U, and the phase margin, ϕ_M of opamp 1. The measurement methods are given in Appendix C. f_U and ϕ_M were obtained from the opamp in a unity gain configuration; the input voltage to it was the dc component V_{CM} plus a sine wave with $100mV$ amplitude. f_U and ϕ_M were measured at V_{CM} between 0.7 and $2.5V$ at every $200mV$. The low frequency gain as a function of V_{CM} is very close to the predicted values obtained from simulations. The load capacitance caused by the measurement set up is about $20pF$, and we can see that the simulation results of f_U with $5pF$ load capacitance are approximately four times the measured values as shown in Table 9.2. The frequency measurements of opamp 1a and opamp 1b are shown in Table 9.3 and Table 9.4, respectively. Both A_{DC} and f_U of opamp 1a stay very flat as a function of V_{CM}. Opamp 1b was tested with $V_b = 1.85V$, and although not as constant as opamp 1a, f_U stays at about $0.5MHz$. Figure 9.9 shows the unity gain frequencies of the three single-stage opamps as a function of the common mode input voltage. The effectiveness of the constant-g_m input stages is clearly seen from this plot.

9.3.3 Step Response

The 2% settling time measurements of opamp 1, connected in the unity gain configuration, obtained from the step response are shown in Table 9.5. Due to the $20pF$ load capacitance which was present at the time of measurement, T_S is accordingly larger compared to the simulation results which were obtained using $5pF$ load. The settling times of opamp 1a and opamp 1b are shown in Table 9.6 and Table 9.7, respectively. Compared to the results of opamp 1, the settling time for the $0.2V$ input step with different initial values are more constant due to more constant f_U. Also, the settling time for larger steps are slower compared to that of opamp 1 since f_U of opamp 1 for the ranges covered by the input steps is always larger than that of opamp 1a and opamp 1b(Figure 9.9.)

9.3.4 Distortion Measurements

Harmonic distortions of the single-stage opamps were measured in the unity gain configuration. The second and the third harmonics and THD of the input signal itself used for the measurements are shown in Table 9.8. In cases where the distortion measurement from the opamp is very small, using Table 9.8, we will be able to determine whether the distortion is caused by the opamp or if it is caused by the input signal itself.

Table 9.2: Experimental results of frequency response of opamp 1 with $20pF$ load.

$V_{CM}(V)$	$A_{DC}(dB)$	$f_U(MHz)$	$\phi_M(^O)$
0.6	41		
0.7	54	0.49	105
0.8	59		
0.9	61	0.51	104
1.0	62		
1.1	64	0.62	105
1.2	66		
1.3	67	0.74	103
1.4	68		
1.5	68	0.84	102
1.6	69		
1.7	70	0.93	104
1.8	70		
1.9	69	0.95	103
2.0	69		
2.1	68	0.95	104
2.2	67		
2.3	65	0.80	103
2.4	62		
2.5	51	0.64	106
2.6	37		
2.7	27		

Table 9.3: Experimental results of frequency response of opamp 1a with $20pF$ load.

$V_{CM}(V)$	$A_{DC}(dB)$	$f_U(MHz)$	$\phi_M(^O)$
0.6	40		
0.7	54	0.48	100
0.8	59		
0.9	61	0.50	99
1.0	62		
1.1	62	0.50	101
1.2	62		
1.3	62	0.52	104
1.4	62		
1.5	62	0.53	106
1.6	62		
1.7	62	0.53	106
1.8	61		
1.9	61	0.53	106
2.0	60		
2.1	60	0.52	105
2.2	60		
2.3	58	0.50	103
2.4	62		
2.5	51	0.49	104
2.6	37		
2.7	28		

Table 9.4: Experimental results of frequency response of opamp 1b with $20pF$ load.

$V_{CM}(V)$	$A_{DC}(dB)$	$f_U(MHz)$	$\phi_M(^O)$
0.6	43		
0.7	55.8	0.50	104
0.8	59.5		
0.9	61	0.50	104
1.0	61		
1.1	62	0.50	105
1.2	62		
1.3	62	0.51	105
1.4	61		
1.5	62	0.51	106
1.6	61		
1.7	61	0.47	103
1.8	61		
1.9	62	0.46	103
2.0	65	0.49	106
2.1	63	0.53	101
2.2	63		
2.3	62	0.50	99
2.4	59		
2.5	48	0.52	100
2.6	35		
2.7	26		

Table 9.5: Experimental results of 2% settling time of opamp 1 with $C_L = 20pF$.

$V_{in\,STEP}(V)$	$V_{in\,INIT}(V)$	$T_{S\uparrow}(\mu sec)$	$T_{S\downarrow}(\mu sec)$
0.2	0.90	1.37	1.50
0.2	1.55	0.82	0.83
0.2	2.20	1.03	0.83
0.4	1.45	0.82	0.91
0.6	1.35	0.79	1.01
0.8	1.25	0.79	1.04
1.0	1.15	0.97	1.23
1.2	1.05	1.09	1.22
1.4	0.95	1.17	1.51
1.6	0.85	1.28	1.63

Table 9.6: Experimental results of 2% settling time of opamp 1a with $C_L = 20pF$.

$V_{in\,STEP}(V)$	$V_{in\,INIT}(V)$	$T_{S\uparrow}(\mu sec)$	$T_{S\downarrow}(\mu sec)$
0.2	0.90	1.34	1.54
0.2	1.55	1.46	1.43
0.2	2.20	1.66	1.73
0.4	1.45	1.57	1.57
0.6	1.35	1.72	1.69
0.8	1.25	1.74	1.67
1.0	1.15	1.85	1.69
1.2	1.05	2.08	1.70
1.4	0.95	2.20	1.94
1.6	0.85	2.33	2.00

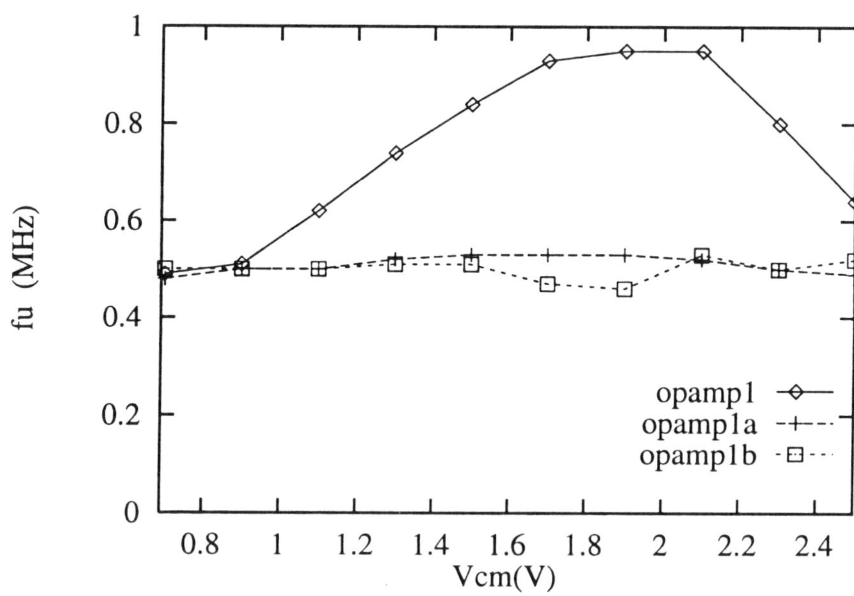

Figure 9.9: The unity gain frequency of the single-stage opamps as a function of V_{CM}.

The measurement results of opamp 1 are shown in Table 9.9. The load resistance due to the input resistance of the device used in the measurement was $1M\Omega$ and it significantly reduces the opamp gain. The output resistance of the single-stage opamps is designed to be more than $30M\Omega$, and $1M\Omega$ load will reduce the low frequency gain by more than $30dB$. Most of the distortion measurements were performed with $10KHz$ input signal, and the effect of the gain reduction at low frequencies will not be as severe as if they were measured at $0.1Hz$, for example. In any case the measured harmonic distortions of opamp 1 as shown in Table 9.9 are larger than the simulation results obtained with no resistive loading. However, since we would like to observe the effect of the constant-g_m input stage on the distortion of opamps, good comparisons can still be made from these measurements.

Measurement results of opamp 1a and opamp 1b are shown in Table 9.10 and Table 9.11, respectively. Note that the entries made with the numbers in bold face types are the ones where the measured distortion was approx-

Table 9.7: Experimental results of 2% settling time of opamp 1b with $C_L = 20pF$.

$V_{in\ STEP}(V)$	$V_{in\ INIT}(V)$	$T_{S\uparrow}(\mu sec)$	$T_{S\downarrow}(\mu sec)$
0.2	0.90	1.55	1.52
0.2	1.55	1.65	1.53
0.2	2.20	1.47	1.42
0.4	1.45	1.68	1.46
0.6	1.35	1.98	1.66
0.8	1.25	1.90	1.54
1.0	1.15	1.98	1.58
1.2	1.05	1.86	1.68
1.4	0.95	1.98	1.76
1.6	0.85	2.00	1.80

Table 9.8: Measured harmonic distortions of 10KHz input signal.

$V_{in}(V)$	$HD2(\%)$	$HD3(\%)$	$THD(\%)$
0.1	0.120	0.010	0.120
0.2	0.055	0.009	0.056
0.3	0.041	0.008	0.042
0.4	0.027	0.007	0.028
0.5	0.024	0.006	0.025
0.6	0.021	0.008	0.023
0.7	0.018	0.007	0.019
0.8	0.017	0.007	0.018
0.9	0.015	0.010	0.018
1.0	0.016	0.006	0.017

imately the same as that of the distortion in the input signal, thus, they correspond to the cases for which the actual distortion of the opamp could be better than the values listed in the tables. We can clearly see the improvements in the distortion performance of the opamps with constant-g_m input stages, both as a function of the input dc voltage, V_{CM}, and as a function of the input signal amplitude, v_{in}. The comparisons are more easily made with the plots shown in Figure 9.10. Figure 9.10(a) shows that when a $10KHz$ sine wave with $0.2V$ amplitude is applied to the opamps, opamp 1a has significantly less distortion for the entire output range. Opamp 1b performs as good as opamp 1a for V_{CM} less than mid-rail; however, be-

Table 9.9: Measured harmonic distortions of opamp 1 with $R_L = 1M\Omega$ and $C_L = 20pF$.

$f_O(Hz)$	$V_{CM}(V)$	$v_{in}(V)$	$HD2(\%)$	$HD3(\%)$	$THD(\%)$
10K	0.9	0.2	0.880	0.084	0.884
10K	1.0	0.2	0.501	0.251	0.560
10K	1.1	0.2	0.295	0.115	0.317
10K	1.2	0.2	0.417	0.023	0.418
10K	1.3	0.2	0.394	0.023	0.395
10K	1.4	0.2	0.335	0.018	0.336
10K	1.5	0.2	0.279	0.012	0.279
10K	1.6	0.2	0.269	0.011	0.269
10K	1.7	0.2	0.269	0.011	0.269
10K	1.8	0.2	0.240	0.009	0.240
10K	1.9	0.2	0.240	0.014	0.240
10K	2.0	0.2	0.347	0.067	0.353
10K	2.1	0.2	0.631	0.104	0.639
10K	2.2	0.2	0.891	0.087	0.895
10K	2.3	0.2	1.135	0.056	1.136
10K	1.65	0.5	0.631	0.061	0.634
10K	1.65	0.6	0.832	0.094	0.837
10K	1.65	0.7	1.023	0.146	1.033
10K	1.65	0.8	1.244	0.251	1.269
10K	1.65	0.9	1.396	0.427	1.460
10K	1.65	1.0	1.445	0.575	1.555
100K	1.0	0.2	1.972	0.316	2.000
100K	1.65	0.2	0.550	0.089	0.557
100K	2.3	0.2	1.799	0.227	1.813

haves similar to opamp 1 when V_{CM} is near $2.0V$. Figure 9.10(b) shows the distortion of the opamps with the dc component of the input voltage fixed at mid-rail while the amplitude of the input signal was increased from $0.5V$ to $1.0V$. Here we can clearly see the superior performance of opamp 1a and opamp 1b over opamp 1.

Table 9.10: Measured harmonic distortions of opamp 1a with $R_L = 1M\Omega$ and $C_L = 20pF$.

$f_O(Hz)$	$V_{CM}(V)$	$v_{in}(V)$	$HD2(\%)$	$HD3(\%)$	$THD(\%)$
10K	0.9	0.2	0.088	0.107	0.139
10K	1.0	0.2	0.197	0.013	0.197
10K	1.1	0.2	0.114	0.037	0.120
10K	1.2	0.2	0.057	0.006	**0.057**
10K	1.3	0.2	0.046	0.008	**0.047**
10K	1.4	0.2	0.048	0.008	**0.049**
10K	1.5	0.2	0.056	0.008	**0.057**
10K	1.6	0.2	0.060	0.008	**0.060**
10K	1.7	0.2	0.065	0.011	0.066
10K	1.8	0.2	0.078	0.014	0.079
10K	1.9	0.2	0.097	0.028	0.101
10K	2.0	0.2	0.162	0.022	0.164
10K	2.1	0.2	0.130	0.049	0.134
10K	2.2	0.2	0.201	0.060	0.210
10K	2.3	0.2	0.380	0.039	0.382
10K	1.65	0.5	0.084	0.031	0.090
10K	1.65	0.6	0.081	0.041	0.091
10K	1.65	0.7	0.116	0.036	0.121
10K	1.65	0.8	0.123	0.058	0.135
10K	1.65	0.9	0.098	0.114	0.150
10K	1.65	1.0	0.118	0.214	0.244
100K	1.0	0.2	0.054	0.0468	**0.071**
100K	1.65	0.2	0.145	0.013	0.146
100K	2.3	0.2	0.708	0.074	0.712

Table 9.11: Measured harmonic distortions of opamp 1b with $R_L = 1M\Omega$ and $C_L = 20pF$.

$f_O(Hz)$	$V_{CM}(V)$	$v_{in}(V)$	$HD2(\%)$	$HD3(\%)$	$THD(\%)$
10K	0.9	0.2	0.090	0.015	0.091
10K	1.0	0.2	0.062	0.014	**0.064**
10K	1.1	0.2	0.059	0.010	**0.060**
10K	1.2	0.2	0.059	0.008	**0.060**
10K	1.3	0.2	0.062	0.012	**0.063**
10K	1.4	0.2	0.099	0.029	0.103
10K	1.5	0.2	0.170	0.037	0.174
10K	1.6	0.2	0.157	0.037	0.161
10K	1.7	0.2	0.170	0.061	0.181
10K	1.8	0.2	0.312	0.054	0.317
10K	1.9	0.2	0.347	0.086	0.356
10K	2.0	0.2	0.232	0.224	0.323
10K	2.1	0.2	0.457	0.224	0.509
10K	2.2	0.2	0.188	0.080	0.204
10K	2.3	0.2	0.048	0.022	**0.053**
10K	1.65	0.5	0.221	0.115	0.249
10K	1.65	0.6	0.178	0.102	0.205
10K	1.65	0.7	0.151	0.100	0.181
10K	1.65	0.8	0.116	0.099	0.152
10K	1.65	0.9	0.126	0.085	0.152
10K	1.65	1.0	0.110	0.070	0.130
100K	1.0	0.2	0.120	0.051	0.130
100K	1.65	0.2	1.148	0.224	1.170
100K	2.3	0.2	0.108	0.010	0.109

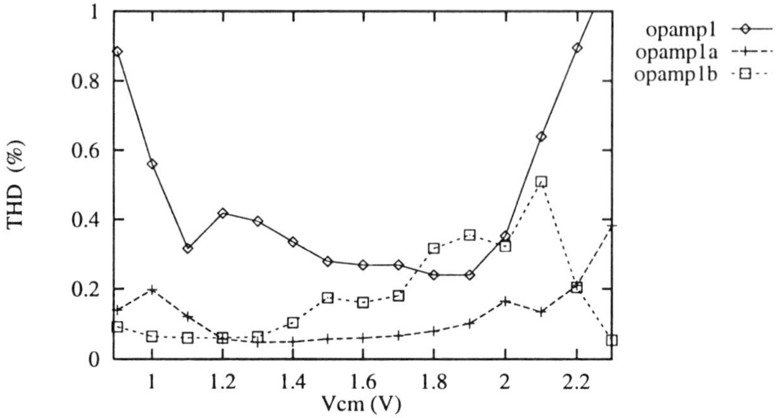

(a)

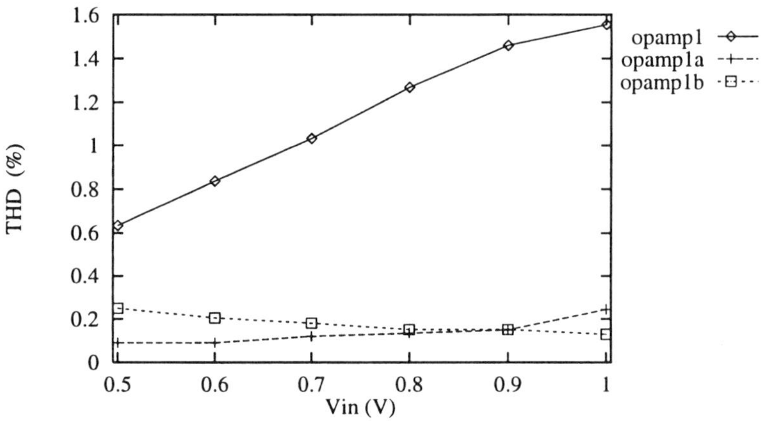

(b)

Figure 9.10: Measured total harmonic distortion of the single-stage opamps: (a) as a function of V_{CM}, $v_{in} = 0.2 \sin 20000\pi t$, (b) as a function of v_{in}, $V_{CM} = 1.65V$.

9.4 Two-Stage Operational Amplifiers

9.4.1 dc Measurements

The dc measurements of the two-stage opamps were obtained using the HP4145. Opamps with single-ended output stages were powered with $V_{DD} - V_{SS} = 3.0V$, and the input stage bias circuit was biased with $I_{nmax} = 18\mu A$. The cascode stage was biased with $I_{C1} = 55\mu A$, and $V_{C3} = 1.6V$, while V_{C2} was 1.6, 1.4, and 1.3V for opamp 2, opamp 2a, and opamp 2b, respectively. The output stage was biased with $I_{min} = 7\mu A$ and $V_{C5} = 2.5V$, and the output node was loaded with $10K\Omega$ resistor. Figure 9.11 shows the measured results of the two-stage opamps. The input-output transfer characteristics of opamp 2 in a unity gain configuration is shown in Figure 9.11(a), and we can see that the rail-to-rail output swing is achieved. The results from opamp 2a and opamp 2b($V_b = 1.7V$) are identical to that of opamp 2. Figure 9.11(b) shows the offset voltage of the two-stage opamps. Offsets are 10 to $20mV$ larger than the single-ended opamps, and are caused by the addition of the output stage.

Figure 9.12 shows the currents flowing in the output transistors of the common source amplifier of opamp 2. Class AB characteristics of the output stage is clearly seen. Since we did not have the pins available for the direct measurements of these currents, we had to extract them by observing the output current, I_O, and the current, I_{C5}, which goes into the V_{C5} node. Since we know that

$$I_{C5} = I_{on} + \frac{1}{2} I_{op} \quad (9.1)$$

$$I_O = I_{op} - I_{on}, \quad (9.2)$$

I_{on} and I_{op} can be found from these two equations, and are given by

$$I_{op} = \frac{2}{3}(I_O + I_{C5}) \quad (9.3)$$

$$I_{on} = \frac{1}{3}(2I_{C5} - I_O). \quad (9.4)$$

Figure 9.13 shows I_{op} and I_{on} of opamp 2a and opamp 2b. The class AB action in opamps 2a and 2b is not as smooth as that of opamp 2; nonetheless they still exhibit the class AB behavior and keep the transistors in the common source amplifiers from turning off.

Figure 9.14 shows the measurement results of opamp 3a in the circuit configuration given in Figure 8.9 with $R = 100K\Omega$ with each output node loaded with $10K\Omega$. Then, we should have

$$V_o = V_{o1} - V_{o2} = -V_{in} = V_{i2} - V_{i1}. \quad (9.5)$$

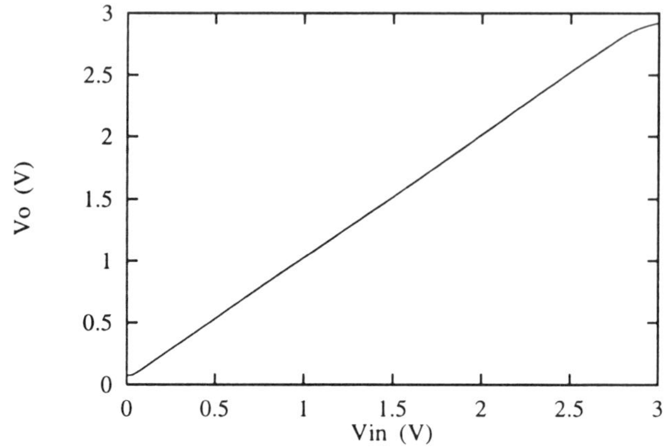

(a)

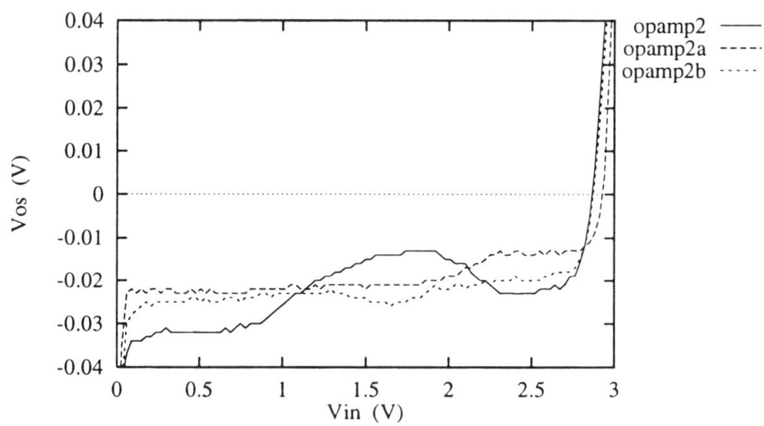

(b)

Figure 9.11: Experimental results of the two-stage opamps in a unity gain configuration. (a) Vin-Vo characteristics. (b) Offset voltages.

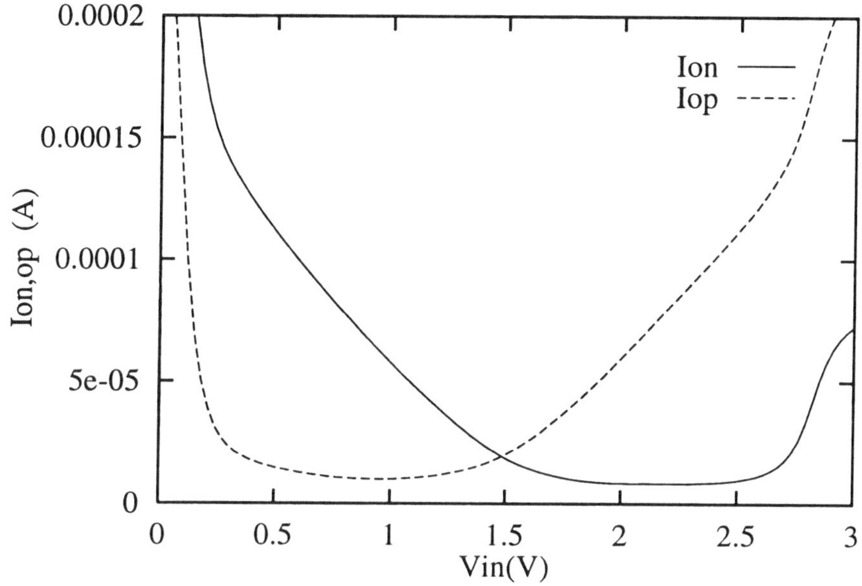

Figure 9.12: Current flow in the transistors in the output stage of opamp 2.

The circuit was powered with $V_{DD} - V_{SS} = 3.3V$. The offset voltage of the fully-differential opamps were smaller with $3.3 - V$ supply than with $3 - V$ supply. Note that the offset voltages of the opamps with single-ended output, both single-stage and two-stage, were not affected by the $0.3V$ difference in the power supply. Both opamp 3a and opamp 3b were biased with $I_{nmax} = 18\mu A$, $V_{C2} = 1.75V$, $V_{C3} = 1.65V$, $I_C = 37\mu A$, $V_{C5} = 2.5V$, and $I_{min} = 7\mu A$. The measurement results of opamp 3b($V_b = 2.2V$) are identical to that of opamp 3a as shown in Figure 9.14 and $\pm 3V$ input and output ranges of the fully-differential opamps are demonstrated.

The offset voltage, V_{OS}, of opamp 3a and opamp 3b are measured as the sum of V_o and V_{in} and it is idealy $0V$. The measured results are shown in Figure 9.15; even though the offset voltages of the single-ended output versions of these opamps were very close to each other, the offset voltage of opamp 3b was measured to be about three times larger than that of opamp 3a. Besides the systematic offset which is caused by the circuit design itself, transistor mismatches are responsible for a large portion of

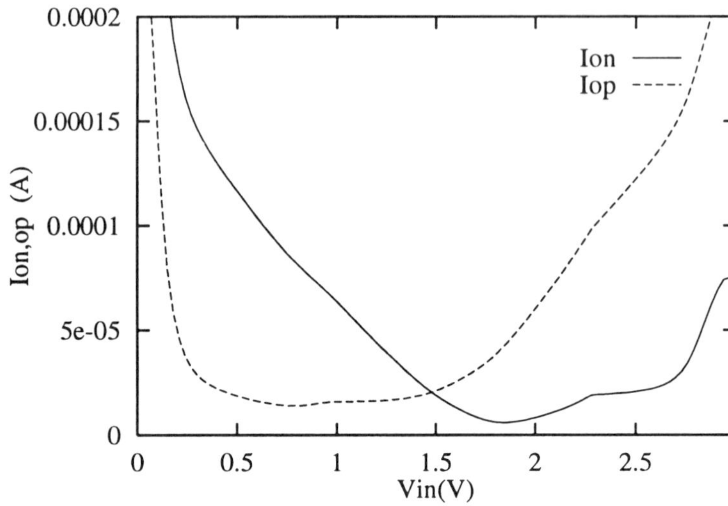

(a)

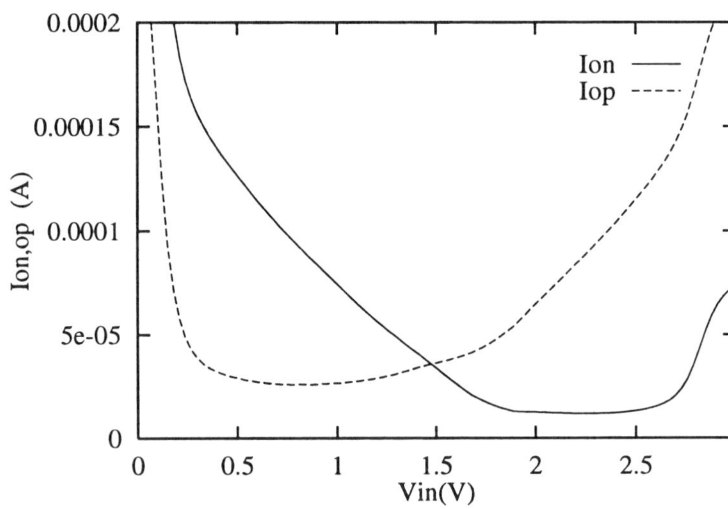

(b)

Figure 9.13: Current flow in the transistors in the output stage of the two-stage opamps. (a) opamp 2a. (b) opamp 2b.

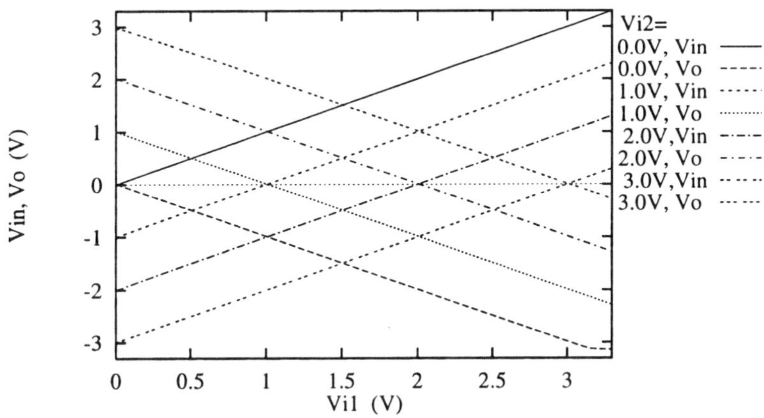

Figure 9.14: dc measurements of the opamp 3a and opamp 3b in the inverting unity gain configuration.

the offset voltages. Due to the fact that mismatches are random in nature, statistical techniques should be used to account for their effects[52, 53].

9.4.2 Frequency Response

Two-stage opamps were powered with $V_{DD} - V_{SS} = 3.3V$ by HP4145 for the same reasons given for the single-stage opamps. Note that while I_{min} for opamp 2a and opamp 2b were externally provided, opamp 2, due to the lack of an extra output pin, has an internal biasing for I_{min} which is equal to $\frac{1}{10}I_{C1}$. All the two-stage opamps were biased with $I_{nmax} = 18\mu A$, $V_{C3} = 1.65V$, $I_{C1} = 37\mu A$, and $V_{C5} = 2.5V$. V_{C2} of opamp 2a and opamp 2b were $1.75V$ while it was $1.85V$ for opamp 2. $I_{min} = 7\mu A$ were provided externally to opamp 2a and opamp 2b while I_{min} of opamp 2 was about $4\mu A(37\mu A/10.)$ These bias conditions were used to obtain the results provided in the rest of this section. Note that results given in the remaining subsections are only of opamps with single-ended output stages; parasitic capacitors present at every pin of the chip significantly affected the frequency and the transient responses. Moreover, the lack of measurement devices which can accurately measure two outputs simultaneously made it difficult to characterize the fully-differential opamps. Thus, based on the similarities between the single-ended and the fully-differential versions, and the results obtained from dc measurements, we can only infer the perfor-

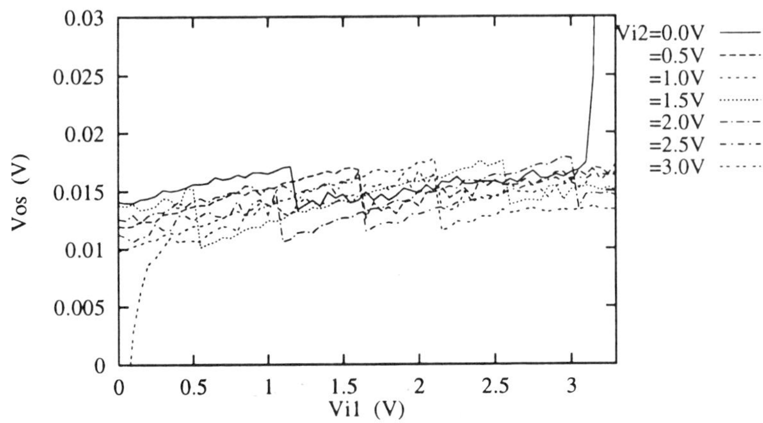

(a)

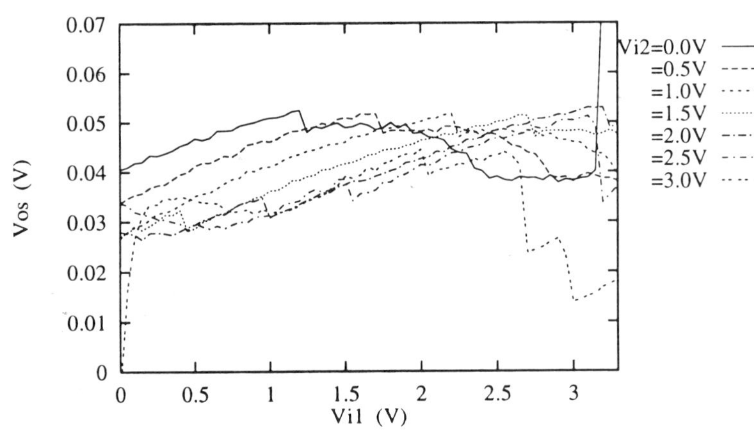

(b)

Figure 9.15: Offset measurements of opamp 3a (a) and opamp 3b (b) in the inverting unity gain configuration.

Table 9.12: Experimental results of frequency response of opamp 2 with $20pF$ load.

$V_{CM}(V)$	$A_{DC}(dB)$	$f_U(MHz)$	$\phi_M(^O)$
0.1	40		
0.3	74	0.90	55
0.5	75	0.92	54
0.7	75	0.92	53
0.9	74	1.03	54
1.1	76	1.25	51
1.3	78	1.50	49
1.5	79	1.63	48
1.7	79	1.73	48
1.9	79	1.76	46
2.1	78	1.76	46
2.3	76	1.73	49
2.5	72	1.40	51
2.7	71	1.19	53
2.9	67	1.19	53
3.1	51	1.15	53

mance of the fully-differential opamps based on the results of the opamps with single-ended output stage.

Tables 9.12, 9.13, and 9.14 show the measured frequency response of the three two-stage opamps with single-ended outputs. We see that the low frequency gain, A_{DC}, of opamp 2 is slightly less than the gain of opamp 2a and opamp 2b. This could be caused by the smaller I_{min} and hence the smaller currents in the output transistors used in opamp 2; however, in theory the gain of the output stage is inversely proportional to the square root of the current and the gain should actually increase with smaller currents. The phase margin of opamp 2 is not as large as the other ones; this is due to the fact that smaller I_{min} caused the second pole of opamp 2 to be smaller than in the other two. More importantly, the phase margin of opamp 2 is the worst when the common mode input voltage is near mid-rail; this is because of the larger unity gain frequency which is closer to the second pole by a factor of two compared to the case where V_{CM} is near one of the rails. Opamps with the constant-g_m input stage definitely have more constant f_U than that of opamp 2 as expected.

Figure 9.16 shows f_U of the three opamps as a function of the common mode input voltage. f_U of opamp 2a and opamp 2b stays almost constant at

Table 9.13: Experimental results of frequency response of opamp 2a with $20pF$ load.

$V_{CM}(V)$	$A_{DC}(dB)$	$f_U(MHz)$	$\phi_M(^O)$
0.1	35		
0.3	83	1.55	92
0.5	87	1.55	87
0.7	88	1.50	75
0.9	90	1.47	74
1.1	88	1.47	72
1.3	90	1.46	72
1.5	88	1.46	72
1.7	88	1.45	70
1.9	83	1.45	70
2.1	83	1.46	65
2.3	82	1.52	59
2.5	78	1.41	53
2.7	78	1.43	64
2.9	74	1.43	64
3.1	52	1.34	63

$1.5MHz$ and $1.4MHz$, respectively, while f_U of opamp 2 changes between $0.9MHz$ and $1.8MHz$. We have successfully demonstrated the effectiveness of the constant-g_m input stages on f_U of both the single and the two-stage opamps. Note that f_U of opamp 2 does not change between $1.5MHz$ and $3MHz$ as if the maximum currents provided to the differential pairs of all the opamps were the same(f_U of opamps 2a and 2b are about $1.5MHz$ and this should be the minimum f_U of opamp 2.) This is because of the transistors M_{2b} and M_{2c}(Figure 8.1) which together carry the same amount of the current as M_2 and similarly because of M_{1b} and M_{1c}. Since M_8(and also M_p) in all the two-stage opamps carries the same amount of current, M_2(also M_{2a}, M_1, and M_{1a}) in opamp 2 carries only two third of the current compared to opamp 2a and opamp 2b.

9.4.3 Step Response

The 2% settling time of the two-stage opamps connected in the unity gain configuration obtained from the step response is shown in Tables 9.15, 9.16, and 9.17. We can see that the settling times are in accordance with the unity gain frequency of opamps for a given common mode input voltage.

Table 9.14: Experimental results of frequency response of opamp 2b with $20pF$ load.

$V_{CM}(V)$	$A_{DC}(dB)$	$f_U(MHz)$	$\phi_M(^O)$
0.1	22		
0.3	86	1.20	65
0.5	85	1.26	63
0.7	87	1.26	65
0.9	87	1.32	65
1.1	88	1.35	65
1.3	85	1.37	65
1.5	83	1.37	63
1.7	84	1.39	64
1.9	81	1.45	57
2.1	78	1.42	60
2.3	78	1.42	61
2.5	76	1.42	60
2.7	75	1.43	59
2.9	71	1.43	60
3.1	57	1.39	59

Note that T_S which is measured with $20pF$ load is only slightly larger than the simulated results which were obtained with $5pF$. This is due to the fact that the output stage buffers the external loading, and as long as the load capacitance is not extremely large, the two-stage opamps will operate with only slight performance degradations.

9.4.4 Distortion Measurements

Results of the distortion measurements of the two-stage opamps are given in Tables 9.18, 9.19, and 9.20. The presence of the $1M\Omega$ load resistance is negligible for these opamps since their output resistance is much less than $1M\Omega$. Smaller A_{DC} may cause larger distortion in opamp 2 at low frequencies compared to opamp 2a and opamp 2b; however, at $10KHz$, which is much larger than the open loop $3-dB$ frequency of the opamp, the open loop gain of the three opamps should be close to each other. Thus, the difference in the distortion performance should be mostly affected by the change in the gain as a function of the common mode input voltage rather than the absolute value of the gain itself. Distortions in the opamps with constant-g_m input stages are generally less than those in opamp 2. Fig-

Table 9.15: Experimental results of 2% settling time of opamp 2 with $R_L = 1M\Omega$ and $C_L = 20pF$.

$V_{in\ STEP}(V)$	$V_{in\ INIT}(V)$	$T_{S\uparrow}(\mu sec)$	$T_{S\downarrow}(\mu sec)$	$P.O_{\uparrow}(\%)$	$P.O_{\downarrow}(\%)$
0.4	0.8	0.82	0.84	16	21
0.4	1.45	0.61	0.55	16	22
0.4	2.1	0.70	0.56	22	20
1.0	1.15	0.86	0.65	7.5	13
1.6	0.85	1.1	0.88	7.5	12
2.4	0.45	1.4	0.90	6.3	7.0

Table 9.16: Experimental results of 2% settling time of opamp 2a with $R_L = 1M\Omega$ and $C_L = 20pF$.

$V_{in\ STEP}(V)$	$V_{in\ INIT}(V)$	$T_{S\uparrow}(\mu sec)$	$T_{S\downarrow}(\mu sec)$	$P.O_{\uparrow}(\%)$	$P.O_{\downarrow}(\%)$
0.4	0.8	0.85	0.74	8.4	11
0.4	1.45	0.91	0.70	10	11
0.4	2.1	0.68	0.60	11	20
1.0	1.15	1.08	0.72	7.5	8.1
1.6	0.85	1.23	0.80	8.4	7.4
2.4	0.45	1.52	0.84	5.6	5.6

Table 9.17: Experimental results of 2% settling time of opamp2b with $R_L = 1M\Omega$ and $C_L = 20pF$.

$V_{in\ STEP}(V)$	$V_{in\ INIT}(V)$	$T_{S\uparrow}(\mu sec)$	$T_{S\downarrow}(\mu sec)$	$P.O_{\uparrow}(\%)$	$P.O_{\downarrow}(\%)$
0.4	0.8	0.77	0.74	12	11
0.4	1.45	0.69	0.73	8.4	8.4
0.4	2.1	0.70	0.67	14	12
1.0	1.15	0.96	0.72	8.1	12
1.6	0.85	1.1	0.61	7.4	8.4
2.4	0.45	1.2	0.72	4.9	2.8

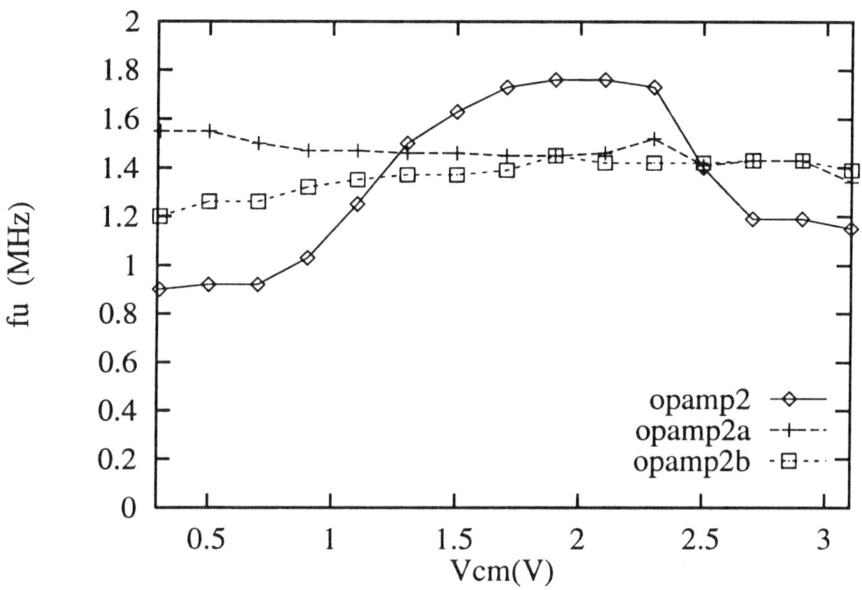

Figure 9.16: The unity gain frequency of the two-stage opamps as a function of V_{CM}.

ure 9.17(a) shows that the input sine wave with $0.2V$ amplitude caused the largest distortion in opamp 2 for almost all the dc component values used. However, for $V_{CM} = 2.6V$, the distortion from opamp 2a was significantly larger than the other two. Over all, opamp 2b seems to perform the best in terms of the distortion level as a function of the dc component of the input voltage. Figure 9.17(b) shows the THD of the opamps as a function of the amplitude of the input voltage where the dc component was fixed at mid-rail. We can clearly see the improvements in the distortion performance resulting from using a constant-g_m input stage in the two-stage opamps.

9.5 Power Up Problem and Solution

We noticed, through measurements, that the input stages with constant-g_m bias circuit does not operate properly depending on the power supply conditions, *i.e.*, the constant-g_m input stage was behaving exactly like that

Table 9.18: Measured harmonic distortions of opamp 2 with $R_L = 1M\Omega$ and $C_L = 20pF$.

$f_O(Hz)$	$V_{CM}(V)$	$v_{in}(V)$	$HD2(\%)$	$HD3(\%)$	$THD(\%)$
10K	0.5	0.2	0.056	0.011	**0.057**
10K	0.8	0.2	0.331	0.041	0.333
10K	1.1	0.2	0.155	0.025	0.157
10K	1.4	0.2	0.129	0.011	0.130
10K	1.7	0.2	0.114	0.010	0.114
10K	2.0	0.2	0.114	0.065	0.131
10K	2.3	0.2	0.160	0.012	0.161
10K	2.6	0.2	0.116	0.011	0.117
10K	2.8	0.2	0.093	0.015	0.094
10K	1.0	0.5	0.178	0.133	0.222
10K	1.65	0.5	0.180	0.029	0.182
10K	2.3	0.5	0.302	0.146	0.336
10K	1.65	1.0	0.484	0.127	0.501
10K	1.65	1.1	0.468	0.135	0.487
10K	1.65	1.2	0.432	0.130	0.451
10K	1.65	1.3	0.389	0.122	0.408
10K	1.65	1.4	0.339	0.110	0.356
10K	1.65	1.5	0.309	0.100	0.325
100K	1.0	0.5	1.059	0.118	1.067
100K	1.65	0.5	0.209	0.127	0.245
100K	2.3	0.5	0.851	0.359	0.924

Table 9.19: Measured harmonic distortions of opamp 2a with $R_L = 1M\Omega$ and $C_L = 20pF$.

$f_O(Hz)$	$V_{CM}(V)$	$v_{in}(V)$	$HD2(\%)$	$HD3(\%)$	$THD(\%)$
10K	0.5	0.2	0.060	0.014	**0.062**
10K	0.8	0.2	0.097	0.034	0.102
10K	1.1	0.2	0.078	0.023	0.081
10K	1.4	0.2	0.054	0.006	**0.054**
10K	1.7	0.2	0.050	0.007	**0.051**
10K	2.0	0.2	0.061	0.023	0.065
10K	2.3	0.2	0.172	0.026	0.174
10K	2.6	0.2	0.462	0.017	0.462
10K	2.8	0.2	0.130	0.043	0.137
10K	1.0	0.5	0.048	0.025	0.054
10K	1.65	0.5	0.035	0.004	0.035
10K	2.3	0.5	0.078	0.119	0.142
10K	1.65	1.0	0.044	0.062	0.076
10K	1.65	1.1	0.054	0.039	0.067
10K	1.65	1.2	0.052	0.022	0.056
10K	1.65	1.3	0.049	0.011	0.050
10K	1.65	1.4	0.044	0.004	0.044
10K	1.65	1.5	0.041	0.010	0.042
100K	1.0	0.5	0.160	0.075	0.177
100K	1.65	0.5	0.125	0.057	0.137
100K	2.3	0.5	0.309	0.543	0.625

Table 9.20: Measured harmonic distortions of opamp 2b with $R_L = 1M\Omega$ and $C_L = 20pF$.

$f_O(Hz)$	$V_{CM}(V)$	$v_{in}(V)$	$HD2(\%)$	$HD3(\%)$	$THD(\%)$
10K	0.5	0.2	0.053	0.007	**0.053**
10K	0.8	0.2	0.053	0.003	**0.053**
10K	1.1	0.2	0.053	0.009	**0.054**
10K	1.4	0.2	0.056	0.015	**0.058**
10K	1.7	0.2	0.114	0.041	0.121
10K	2.0	0.2	0.053	0.100	0.113
10K	2.3	0.2	0.053	0.009	**0.054**
10K	2.6	0.2	0.053	0.006	**0.053**
10K	2.8	0.2	0.053	0.007	**0.053**
10K	1.0	0.5	0.024	0.005	**0.025**
10K	1.65	0.5	0.046	0.068	0.082
10K	2.3	0.5	0.054	0.013	0.055
10K	1.65	1.0	0.026	0.048	0.055
10K	1.65	1.1	0.022	0.049	0.054
10K	1.65	1.2	0.019	0.045	0.049
10K	1.65	1.3	0.020	0.043	0.048
10K	1.65	1.4	0.018	0.039	0.043
10K	1.65	1.5	0.017	0.036	0.040
100K	1.0	0.5	0.068	0.065	0.094
100K	1.65	0.5	0.070	0.331	0.338
100K	2.3	0.5	0.145	0.279	0.314

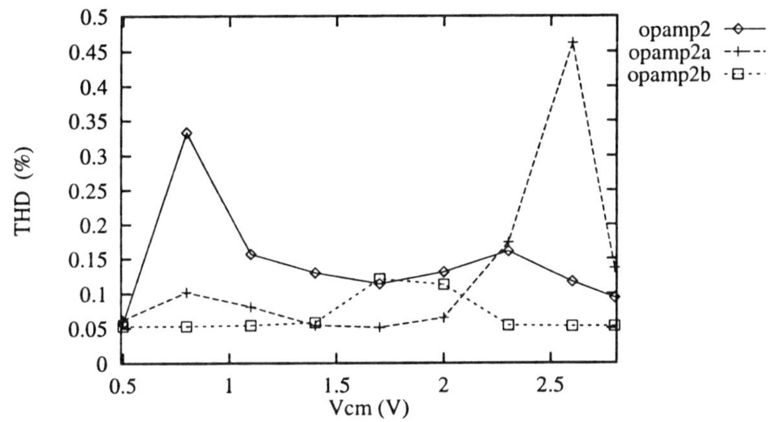

(a)

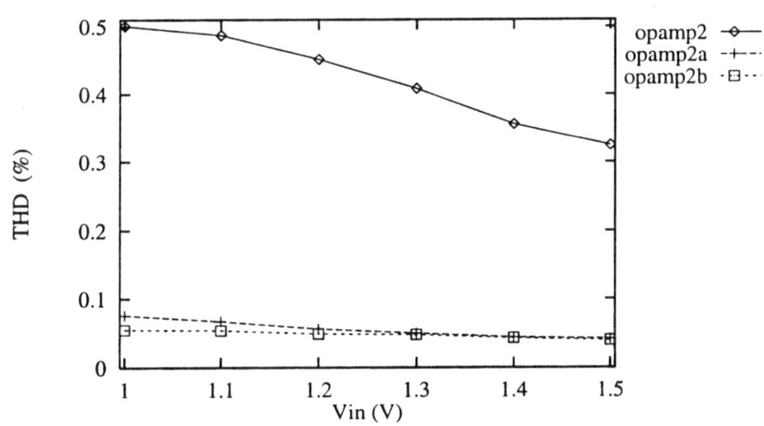

(b)

Figure 9.17: Measured total harmonic distortion of the two-stage opamps: (a) as a function of V_{CM}, $v_{in} = 0.2 \sin 20000\pi t$, (b) as a function of v_{in}, $V_{CM} = 1.65V$.

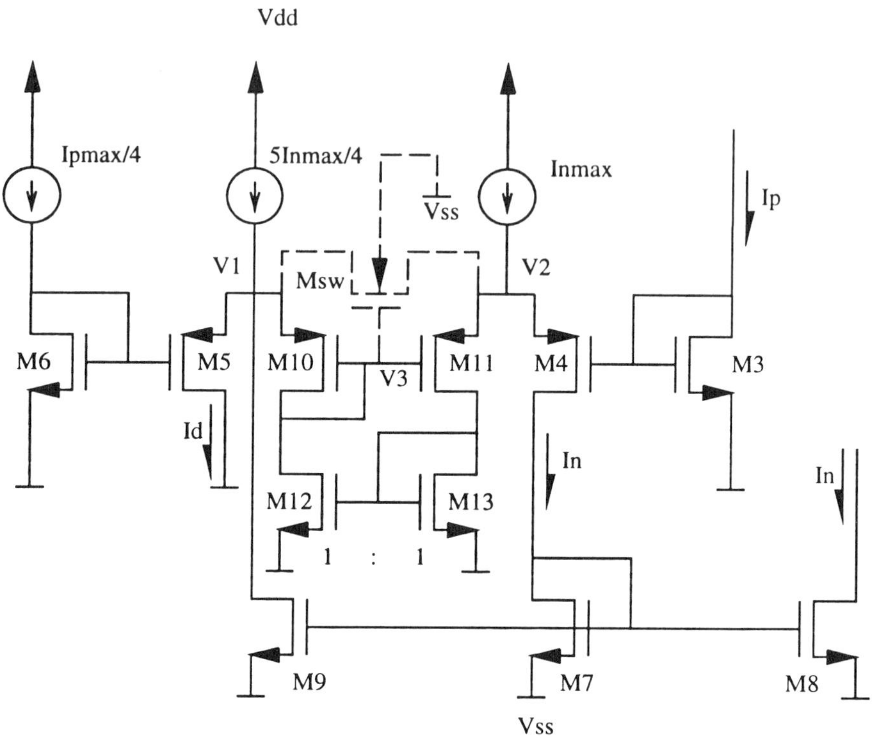

Figure 9.18: Constant-g_m bias circuit with M_{sw} added to prevent the transient problem at the power up.

of the input stage without the constant-g_m bias circuits. This implies that I_n is completely independent of I_p.

Figure 9.18 shows the part of the constant-g_m bias circuit similar to that of Figure 5.5. We suspect that the problem is in the transient response of the circuit consisting of $M_{10} - M_{13}$, and the simulation results supporting this observation will be provided along with a possible solution to this problem. Ignoring the n-channel transistor, M_{sw}, for the moment, the circuit shown in Figure 9.18 is the constant-g_m bias circuit used in the silicon implementations of the low-voltage opamps. In order for I_n to be independent of I_p as we sometimes observed in the experiments, the current flowing in M_4, I_n, must be constant. The purpose of the transistors $M_{10} - M_{13}$ was to keep V_2 constant and equal to V_1; however, if I_n is a constant, then V_2 must be independent of V_1, and this implies that $M_{10} - M_{13}$ are not operating at all.

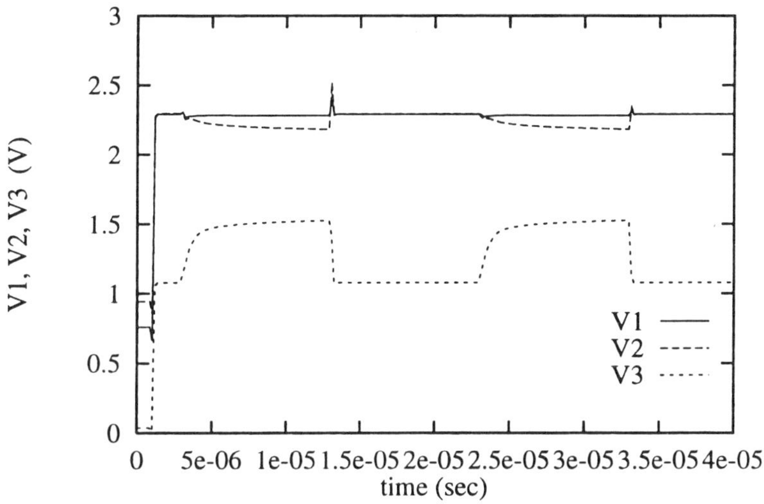

(a)

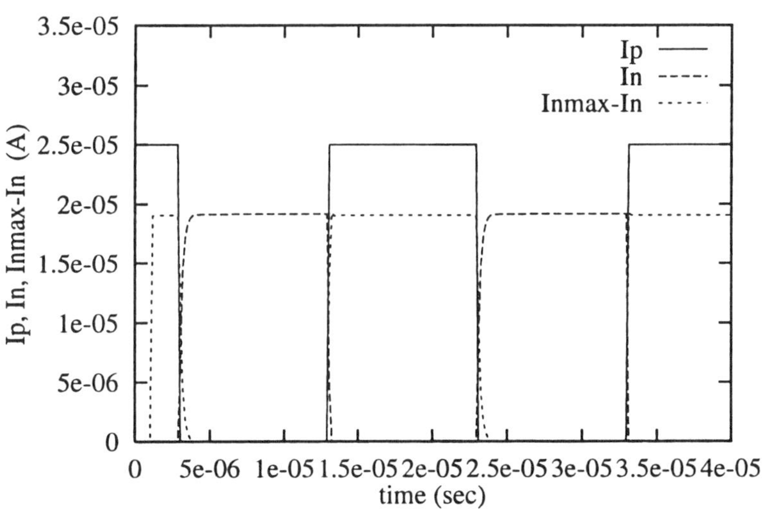

(b)

Figure 9.19: Transient response of the constant-g_m bias circuit with V_{DD} fixed at $3V$.

The only time that the circuit of $M_{10} - M_{13}$ is not operating is when the current flowing in it is zero, and this is supposed to take place only when I_p is zero and I_n has reached I_{nmax}; normally, V_2 should still stay close to V_1 and as soon as I_p becomes nonzero, M_{11} starts conducting and the circuit operates as it should. Figure 9.19 shows the simulation results of the Figure 9.18(without M_{sw}) with $V_{DD} - V_{SS}$ being fixed at $3V$. I_p was stepped between 0 and $25\mu A$, and we can see that $I_n = I_{nmax} = 19\mu A$ when $I_p = 0$ and $I_n = 0$ when I_p goes up to $25\mu A$. This is, of course, because V_2 is following V_1 when I_p is large, and staying close to V_1 when $I_n = I_{nmax}$.

Figure 9.20 shows the simulation results of the same circuit except that V_{DD} is now ramped from 0 to $3V$ at $1\mu sec$ while $V_{SS} = 0$. We see that V_2 is larger than V_1 when $I_p = 25\mu A$, large enough to enable both M_3 and M_4 to conduct large amounts of current. $I_{nmax} - I_n$ is very small regardless of I_p and forces V_3 to closely follow V_2. This problem is caused by the fact that $M_{10} - M_{13}$ are completely turned off as the rest of the circuit is when $V_{DD} = 0$, and never being able to turn on while other transistors start to conduct current as V_{DD} reaches $3V$. Note that if a very small part of I_{nmax} connected to M_{11} and M_4 flows into M_{11} at the power up time, the circuit will operate properly, and this is why we were able to test the opamps and obtain good results after finding a good set of values for V_{DD} and V_{SS}.

Observing the simulation results given in Figure 9.20(a), we notice that when the circuit is not operating properly, V_3 is much larger than its expected value, and moreover it becomes larger than V_1 and/or V_2. Thus, placing a switch, which turns on to short V_1 and V_2 when there is a problem while completely being off normally, should solve the transient problem associated with the power up. Even though the NMOS switch M_{sw} shown in Figure 9.18 does not turn on fully, its presence is sufficient to provide $M_{10} - M_{13}$ with some current such that they are not completely and indefinitely turned off. The simulation results are shown in Figure 9.21; V_2 and V_3 starts to become larger than expected at the beginning, but they are eventually brought down to where they are expected to be, and a correct circuit operation is resumed.

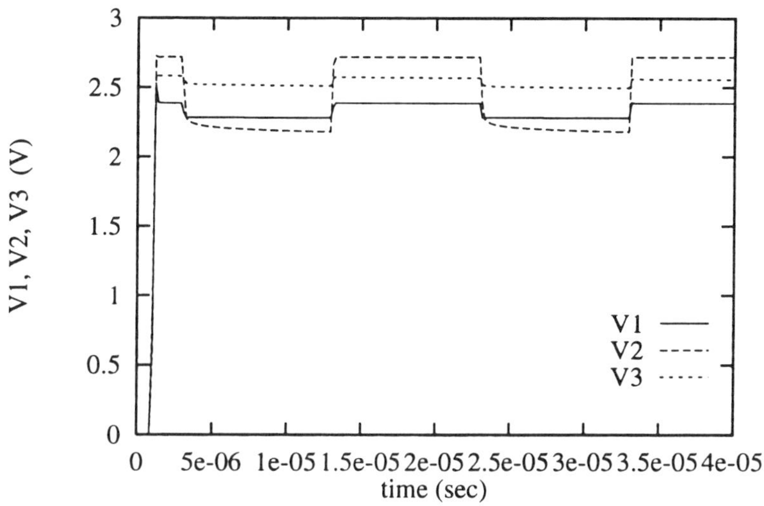

(a)

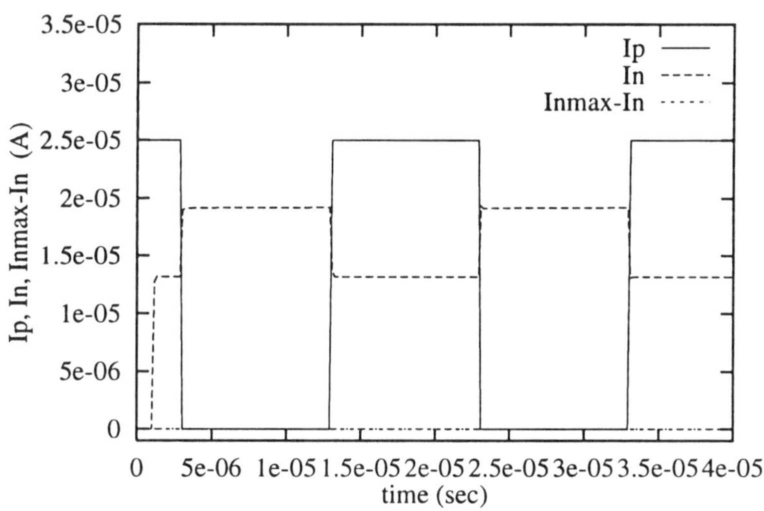

(b)

Figure 9.20: Transient response of the constant-g_m bias circuit with ramped V_{DD}.

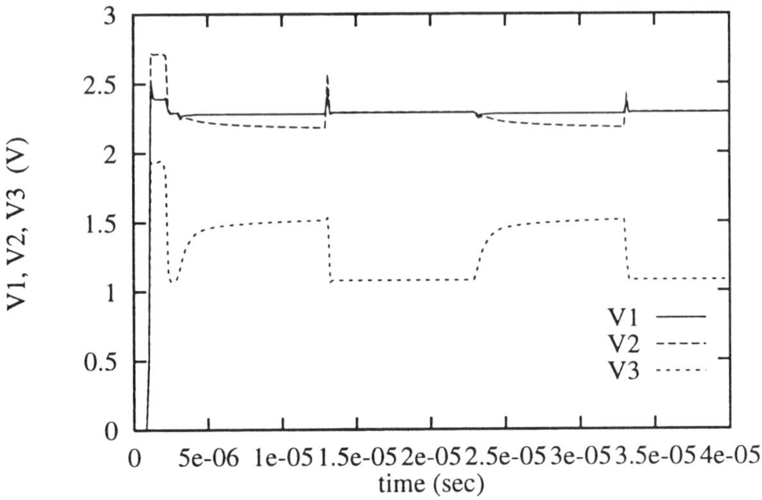

(a)

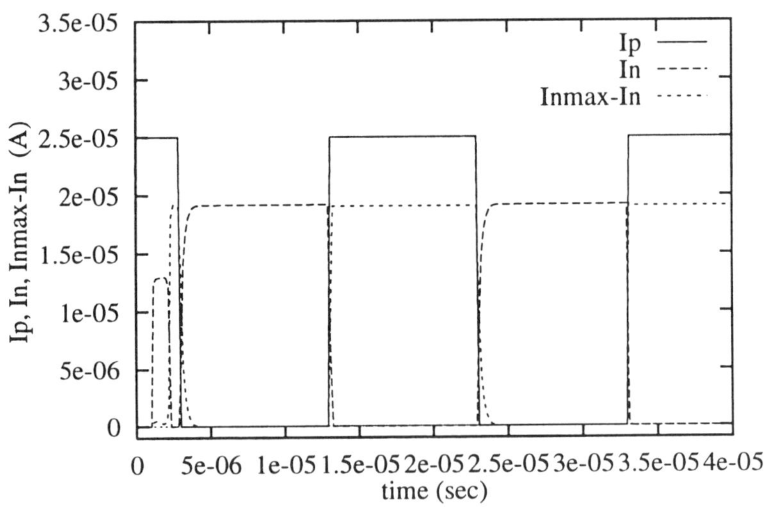

(b)

Figure 9.21: Transient response of the constant-g_m bias circuit with M_{sw} added and with ramped V_{DD}.

Chapter 10

Conclusion and Future Work

In the ever progressing world of the semiconductor business, transistor scaling means more profit to some people and it means more challenges to others. As members of the group who accepted such challenges, we have designed MOS transistor circuits which overcome the performance degradations derived from the scaling, and successfully implemented them on a silicon.

In Chapter I, the necessity to reduce the power supply voltages to 3.3V for CMOS VLSI circuits was pointed out and hence the motivation for developing rail-to-rail operational amplifiers with constant-g_m input stage was set.

In Chapter II, some traditional analog circuits such as a differential pair and a source follower were analyzed, and the reduction in the operating range due to the power supply reduction was demonstrated. It was pointed out that the circuits operating with a 3-V supply must have rail-to-rail input and output range, that is, they must utilize the entire 3V range of operation. Such circuits can be implemented using n- and p-channel differential pairs in parallel as an input stage and a common source amplifier as an output stage. However; it was shown that the use of two differential pairs results in an input transconductance, g_{mT}, which is a strong function of the common mode input voltage, and introduces some unwanted side effects. One of which is the dependence of g_{mT} on the unity gain frequency, f_U, of an opamp; this complicates the frequency compensation of the opamp and results in a tradeoff between the unity gain frequency and the phase margin. It was also demonstrated that the non-constant-g_{mT} which results in a non-constant opamp gain introduces harmonic distortions. Thus, in Chapter II, the motivation for designing a CMOS input stage with a constant-g_m input stage was stated.

In Chapter III, some of the recently reported CMOS input stages with constant-g_m characteristics were given and the operation of the circuits were explained. However, the whole idea was based on an assumption that K_n of an n-channel transistor can be matched with K_p of a p-channel transistor by correctly selecting the aspect ratios of the two types of transistors. The variation in the actual values of K_n and K_p from one processing run to the next processing run was pointed out using parameter sets collected from MOSIS. Thus, the main goal of the research was established at this point; to design a rail-to-rail CMOS input stage with constant-g_m without the requirement of matching n- and p-channel transistors.

In Chapter IV, new circuit techniques which can be used as bias circuits for constant-g_m input stage without the necessity for matching n- and p-channel transistors were introduced. The bias circuits take in the current, I_p, flowing in the p-channel differential pair as an input current and deliver I_n which is to be used as the bias current for the n-channel differential pair. Two methods for directing I_p into the constant-g_m bias circuits were introduced.

In Chapter V, the circuit blocks introduced in Chapter IV were brought together to form an input stage; however, due to the weak inversion operation of some transistors, the input stage did not operate as expected. I_n was allowed to be much larger than expected by the design. The solution to this problem was provided through an addition of a CMOS circuit which prevents the problem by limiting I_n to its desired maximum value while preserving the desired properties of the original circuit. Simulation results of the constant-g_m input stages exhibited g_{mT} as a function of V_{CM} which were much more constant than the input stage without a constant-g_m circuit. However; the curves were rather "bumpy", and this led us to investigate effect of the transistor nonidealities on the constant-g_m input stages. In the design, transconductances of the input differential pairs were assumed to be proportional to the square root of the drain current; however, when one of the differential pairs is conducting a very small amount of current, transistors in that pair are in the weak inversion region and the transconductance is proportional to the current itself. This of course introduces a small variation in g_{mT}. Note that when one of the pairs is in weak inversion, the other pair is conducting a lot of current and the contribution of the transconductance of that pair dominates g_{mT}, thus the deviation is kept to only a few percent. Even when all the transistors are operating in the saturation region in strong inversion, the input stage exhibits nonidealities due to mobility degradation and the body effect. The circuits were analyzed taking these effects into account, and results indicated an addition of a few percent in the deviation.

In Chapter VI, opamp architectures suitable for design with the newly developed input stages are presented. Existing rail-to-rail output stages with class AB control were given and the necessity for modifications were pointed out. A modified output stage and a class AB control circuit were introduced and their operations were verified by computer simulation.

In Chapter VII, three single-stage opamps with rail-to-rail input range were designed. Two of them had constant-g_m input stages while one did not. Two-stage versions of the opamps were designed in Chapter VIII which also contained the design of two fully-differential opamps with constant-g_m input stage. Extensive simulation results were provided in each chapter.

Measured results of the opamps fabricated in a MOSIS $2\mu m$ p-well process were given in Chapter IX. The results were as expected for the most part. Constant-g_m input stage exhibited approximately 10% deviation as the common mode input was swept between $V_{SS} = 0$ and $V_{DD} = 3V$. The unity gain frequency of the opamps with constant-g_m input stages was accordingly constant. Measured THD of the opamps with constant-g_m input stage was significantly better than those without it. Thus, we have successfully designed and implemented rail-to-rail opamps with a constant-g_m input stage.

10.1 Future Work

Since the work presented in this book represent only the beginning of a new generation of low-voltage analog circuits, there is still plenty of work to be followed. Analog circuits which need to be developed in order to be useful for a 3-V power supply include: linear voltage to current converters, four-quadrant multipliers, and active tunable resistor circuits. The development of these low-voltage circuits will be very useful to next generation analog and mixed analog/digital VLSI signal processing applications. All of these circuits operate in an open loop configuration and the task of maintaining the transconductance or the resistance values constant, for a rail-to-rail operation in the input stage, is extremely difficult. This is because the circuit parameters not only have to be constant for the change in the common mode input voltage, but also have to remain constant for the change in the differential input voltage which is for all practical purpose zero in the case of opamps.

Design of above mentioned circuits needs to be carried out carefully, and we believe that the work will go beyond described here. In the following we will suggest some tasks which could be seen as a natural extension to the content of this work. Needless to say, more efforts should be made toward the development of analog MOS IC's suitable for operation from lower supply voltages, *e.g.* $2.7V, 1.8V$ and down to $0.9V$. Results of such efforts start to appear in the literature[62].

10.1.1 Improving the Opamp Performance

Since the main goal of the research was to design and implement the constant-g_m input stage and to demonstrate its effectiveness, no major effort was given to optimize some of the opamp performances. In fact, the unity gain frequency of 1 to $2MHz$ was chosen so that the opamps can be tested without major difficulty. We have successfully demonstrated the constant f_U as a function of the common mode voltage; however, usefulness of such characteristics should be demonstrated when opamps are designed to have the maximum f_U possible for a given process. Without the constant f_U, one would have to choose either the smaller f_U or the smaller phase margin for a given location of the second pole. We did observe some of these behaviors from our opamps, but the second pole of the opamp without the constant-g_m input stage could have been designed to be located at a higher frequency than that of the opamps with constant-g_m input stage so that a good phase margin can be maintained for the entire common mode input range.

The device sizes of the transistors were chosen to make sure the circuits function properly. This sometimes resulted in making the transistors much larger than they were required to be. That is, no optimization in terms of the circuit layout area was performed, and this should certainly be implemented at some point in the future.

In the previous chapter, the power up problem was investigated and a simple solution was given. Even though the simulation results support the effectiveness of the given solution, further investigation is certainly required. First, it should be analyzed more carefully, to make sure that the solution is valid. Since the problem takes place in the transient at the power up where the current flow in the transistors is very small, the accuracy of the simulation must be considered. Second, since the circuit that has the power up problem itself was an additional circuit to the constant-g_m bias circuit in order to overcome problems associated with the weak inversion region operation, the solution to prevent the power up problem was actually a solution to a solution. This means that it may be possible to come up with a circuit which solves both the weak inversion and the power up problem at the same time. Thus, some effort should be given to develop such a circuit.

10.1.2 Offset Voltages

In the measurement results, we observed that the dc offset voltage of the opamps was changing as a function of the common mode voltage. Note that the offsets are introduced due to the mismatches(such as V_T) between the two transistors of the same type in the differential pairs. In opamps which have only one differential pair at the input stage, the offset voltage is constant with its common mode range. However, an opamp with an input stage which has n- and p-differential pairs in parallel will have a total offset voltage which is the sum of the offset voltages from each differential pair, and the total offset voltage will be different when both pairs are operating compared to the case when only one of the pairs is operating. When a large common mode voltage swing is present at the opamp input terminals, the variation in the offset voltage will appear as a small signal and may result in an error in signal processing.

Analyzing the effect of the common mode dependent offset voltage is especially important when the power supply voltage is reduced. Note that since the mismatch in transistor parameters is a random process, the sum of two such effects will be random also, and statistical analysis is needed to investigate the offset voltages. Such studies may lead to discovery of an optimal layout pattern for the differential pairs in rail-to-rail input stages in order to minimize the absolute value of the offset voltage, as well as the variation of it as a function of the common mode voltage.

Appendix A

MOSIS $2\mu m$ P-well Process Parameters

Four sets of process parameters from MOSIS $2\mu m$ p-well process are given in the following. They were obtained from two different runs and both the BSIM and the LEVEL 2 parameters are provided. The first run is referred to as N35S; parameters from this set were used for the design of the constant-g_m input stages and the opamps. The second process run is referred to as N3CM and this is the process that was used to fabricate the circuits in this book.

A.1 BSIM Parameters for N35S

```
.MODEL NM NMOS(LEVEL=4
+ VFB -0.796172     LVFB 3.29555E-02    WVFB -4.08929E-2
+ PHI 0.759208      LPHI 1.23252E-25    WPHI 4.84016E-25
+ K1 1.07084        LK1  -5.67365E-3    WK1  4.13282E-1
+ K2 1.29926E-3     LK2  5.08503E-2     WK2  2.73835E-2
+ ETA -1.46013E-3   LETA 8.04517E-3     WETA 3.77627E-3
+ MUZ 5.99836E+2    DL 6.0685E-1        DW 2.30223E-1
+ U0 5.17285E-2     LU0  5.18907E-2     WU0   -4.34961E-2
+ U1 1.72381E-2     LU1  6.3391E-1      WU1   -2.36149E-1
+ X2M  1.58108E+1   LX2M -3.33769E+1    WX2M   4.34124E+1
+ X2E 1.775E-4      LX2E  -4.09641E-3   WX2E -3.24543E-3
+ X3E 9.79788E-5    LX3E -4.04993E-4    WX3E    -3.8703E-3
+ X2U0 2.98586E-3   LX2U0 -1.35627E-2   WX2U0     2.0798E-2
+ X2U1 -1.18458E-3  LX2U1 -9.89964E-3   WX2U1     2.68566E-2
+ MUS 5.60395E+2    LMUS 3.85123E+2     WMUS 1.76679E+1
```

```
+ X2MS 8.61942E-0  LX2MS  -4.38190E+1 WX2MS 8.81348E+1
+ X3MS -6.56653    LX3MS  7.29079E+1  WX3MS -1.57345E+1
+ X3U1 -1.18345E-2 LX3U1  8.13926E-2  WX3U1 -1.01601E-2
+ TOX=0.0394 VDD=5 DL0=0 DW0=0 AD=0
+ PD=0 AS=0 PS=0 CGDO=3.98897E-10 CGSO=3.98897E-10
+ CGBO=4.03241E-10
+ N0=1 NB0=0 ND0=0 RSH=0 CJ=0 CJW=0 DS=0 WDF=0 XPART=1
+ IJS=0 JSW=0 PJ=0.7 PJW=0.7 MJ=0 MJW=0)

.MODEL P1 PMOS(LEVEL=4
+ VFB -0.470843    LVFB 3.58469E-1    WVFB 5.10888E-1
+ PHI 6.02571E-1   LPHI 2.53717E-25   WPHI -9.24526E-25
+ K1 8.31376E-1    LK1 -4.92295E-1    WK1 -5.25193E-1
+ K2 1.0684E-1     LK2 -9.67435E-2    WK2 -2.0141E-1
+ ETA -1.46881E-2  LETA 6.35594E-2    WETA -3.75992E-3
+ MUZ 2.27366E+2   DL 5.68813E-1      DW  3.34938E-1
+ U0 0.11856       LU0 0.0435904      WU0 -0.0789476
+ U1 3.22507E-2    LU1 2.59425E-1     WU1 -0.0705238
+ X2M 9.16843      LX2M -3.70129      WX2M 5.90459
+ X2E -2.10312E-3  LX2E -5.76318E-4   WX2E  -5.52162E-4
+ X3E 7.67023E-4   LX3E -8.08085E-4   WX3E -8.59729E-3
+ X2U0 5.4276E-3   LX2U0 -2.145E-3    WX2U0 3.54882E-3
+ X2U1 1.44500E-4  LX2U1  5.01708E-3  WX2U1 5.80759E-3
+ MUS 264.361      LMUS 9.27782E+1    WMUS -3.17705E+1
+ X2MS 1.07391E+1  LX2MS -9.81468E-1  WX2MS 6.79508
+ X3MS -0.182785   LX3MS  1.07187E+1  WX3MS  6.0472
+ X3U1 -1.64764E-2 LX3U1  -1.27817E-3 WX3U1  0.0225465
+ TOX=0.0394 VDD=5 DL0=0 DW0=0 AD=0
+ PD=0 AS=0 PS=0 CGDO=3.73895E-10 CGSO=3.73895E-10
+ CGBO=4.29588E-10
+ N0=1 NB0=0 ND0=0 RSH=0 CJ=0 CJW=0 DS=0 WDF=0 XPART=1
+ IJS=0 JSW=0 PJ=0.7 PJW=0.7 MJ=0 MJW=0)
```

A.2 LEVEL 2 Parameters for N35S

```
.MODEL NM NMOS LEVEL=2 LD=0.2045U TOX=394.00000E-10
+ NSUB=2.174E+16 VTO=0.8819 KP=5.081000E-05 GAMMA=0.9693
+ PHI=0.6 U0=579.8 UEXP=0.1531 UCRIT=81740
+ DELTA=7.67 VMAX=66140.0 XJ=0.200000U LAMBDA=2.2660E-02
+ NFS=3.91E+11 NEFF=1 TPG=1.000000
+ RSH=21.830000 CGDO=2.6885E-10 CGSO=2.6885E-10
+ CGBO=3.8386E-10
+ CJ=3.9770E-04 MJ=0.4410 CJSW=4.2372E-10 MJSW=0.338141
+ PB=0.800000
```

```
* Weff = Wdrawn - delta W
* The suggested delta W is -0.2332 um

.MODEL P1 PMOS LEVEL=2 LD=0.2637U TOX=394.000008E-10
+ NSUB=6.803E+15 VTO=-0.7613 KP=1.8019E-05 GAMMA=0.5422
+ PHI=0.6 UO=205.6 UEXP=0.3569 UCRIT=98800
+ DELTA=3.331 VMAX=999900 XJ=0.200000U LAMBDA=4.612000E-02
+ NFS=3.230000E+11 NEFF=1.000 TPG=-1.000000
+ RSH=70.780000 CGDO=3.4667E-10 CGSO=3.4667E-10
+ CGBO=3.6132E-10
+ CJ=2.0787E-04 MJ=0.4926 CJSW=1.7646000E-10 MJSW=0.049688
+ PB=0.800000
* Weff = Wdrawn - delta W
* The suggested delta  is 0.0636 um
```

A.3 BSIM Parameters for N3CM

```
.MODEL NM NMOS(LEVEL=4
+ VFB -0.83213      LVFB 2.65049E-02   WVFB -1.17185E-1
+ PHI 0.775819      LPHI 0             WPHI 0
+ K1 1.1249         LK1  1.55206E-1    WK1  6.77539E-1
+ K2 -1.0957E-2     LK2  1.131197E-1   WK2  9.77784E-2
+ ETA -4.55825E-3   LETA 1.61677E-2    WETA 1.11654E-2
+ MUZ 5.77210E+2    DL 7.33504E-1      DW -6.69874E-2
+ U0 5.41079-2      LU0  5.89261E-2    WU0   -4.63826E-2
+ U1 1.51896E-2     LU1  6.18707E-1    WU1   -1.86506E-1
+ X2M  1.40388E+1   LX2M -3.10674E+1   WX2M   5.64016E+1
+ X2E -2.95684E-4   LX2E  -7.98743E-3  WX2E -4.25596E-3
+ X3E 1.5693E-4     LX3E -5.32536E-5   WX3E    -1.13143E-2
+ X2U0 1.87192E-3   LX2U0 -1.35851E-2  WX2U0    3.1422E-2
+ X2U1 2.17438E-3   LX2U1 7.38676E-3   WX2U1    -9.99812E-4
+ MUS 5.40454E+2    LMUS 4.46351E+2    WMUS -5.0147E+1
+ X2MS 6.28862E+0   LX2MS -2.30657E+1  WX2MS 1.03264E+2
+ X3MS -5.04894     LX3MS  6.5474E+1   WX3MS  9.79183
+ X3U1 -3.81874E-3 LX3U1  -6.59205E-2 WX3U1  0.0171228
+ TOX=0.0403 VDD=5 DL0=0 DW0=0 AD=0
+ PD=0 AS=0 PS=0 CGDO=4.71382E-10 CGSO=4.71382E-10
+ CGBO=3.56273E-10
+ N0=1 NB0=0 ND0=0 RSH=0 CJ=0 CJW=0 DS=0 WDF=0 XPART=1
+ IJS=0 JSW=0 PJ=0.7 PJW=0.7 MJ=0 MJW=0)
```

```
.MODEL P1 PMOS(LEVEL=4
+ VFB -0.353528    LVFB 7.86E-2       WVFB 5.98371E-1
+ PHI 7.05006E-1   LPHI -2.40275E-24  WPHI 4.46805E-24
+ K1 4.82671E-1    LK1 -1.77006E-1    WK1 9.6567E-1
+ K2 1.131177E-2   LK2 -1.03689E-2    WK2 2.45733E-1
+ ETA -1.03918E-2  LETA 6.82136E-2    WETA 2.09421E-2
+ MUZ 2.3729E+2    DL 3.21846E-1      DW  -1.67273E-1
+ U0 0.11624       LU0 0.056091       WU0 -0.0575726
+ U1 1.65663E-1    LU1 1.19191E-1     WU1 -3.74876E-2
+ X2M 10.3841      LX2M -4.57516      WX2M 4.193
+ X2E 8.65489E-4   LX2E -5.82212E-3   WX2E  -6.65438E-3
+ X3E 2.0378E-3    LX3E -4.28931E-3   WX3E -1.0456E-2
+ X2U0 7.74901E-3  LX2U0 -4.61316E-3  WX2U0 1.98612E-3
+ X2U1 3.93253E-3  LX2U1  2.62849E-3  WX2U1 1.68117E-2
+ MUS 236.397      LMUS 2.10762E+2    WMUS -1.99837E+1
+ X2MS 9.54356E+0  LX2MS 1.26727E+0   WX2MS 1.76634E+1
+ X3MS -2.50174    LX3MS  1.87453E+1  WX3MS  12.2179
+ X3U1 -1.04744E-1 LX3U1  1.65484E-1  WX3U1  -7.02258E-3
+ TOX=0.0403 VDD=5 DL0=0 DW0=0 AD=0
+ PD=0 AS=0 PS=0 CGDO=2.06833E-10 CGSO=2.06833E-10
+ CGBO=3.72679E-10
+ N0=1 NB0=0 ND0=0 RSH=0 CJ=0 CJW=0 DS=0 WDF=0 XPART=1
+ IJS=0 JSW=0 PJ=0.7 PJW=0.7 MJ=0 MJW=0)
```

A.4 LEVEL 2 Parameters for N3CM

```
.MODEL NM NMOS LEVEL=2 LD=0.2908U TOX=408.00000E-10
+ NSUB=2.329E+16 VTO=0.9537 KP=4.8918000E-05 GAMMA=1.0262
+ PHI=0.6 U0=570.9 UEXP=0.1771 UCRIT=78200
+ DELTA=8.969 VMAX=62920.0 XJ=0.200000U LAMBDA=2.7720E-02
+ NFS=3.91E+11 NEFF=1 TPG=1.000000
+ RSH=20.870000 CGDO=3.7376E-10 CGSO=3.7376E-10
+ CGBO=4.135E-10
+ CJ=3.5619E-04 MJ=0.4268 CJSW=5.8074E-10 MJSW=0.388144
+ PB=0.800000
* Weff = Wdrawn - delta W
* The suggested delta W is -0.2332 um

.MODEL P1 PMOS LEVEL=2 LD=0.2325U TOX=408.000008E-10
+ NSUB=5.319E+15 VTO=-0.7536 KP=2.077E-05 GAMMA=0.4904
+ PHI=0.6 U0=242.4 UEXP=0.3484 UCRIT=63670
+ DELTA=2.819 VMAX=999900 XJ=0.200000U LAMBDA=5.775000E-02
+ NFS=3.230000E+11 NEFF=1.000 TPG=-1.000000
```

```
+ RSH=66.0000 CGDO=2.9883E-10 CGSO=2.9883E-10
+ CGBO=4.1839E-10
+ CJ=1.7115E-04 MJ=0.4787 CJSW=1.4918000E-10 MJSW=-0.101996
+ PB=0.700000
* Weff = Wdrawn - delta W
* The suggested delta  is 0.0636 um
```

Appendix B

Circuit Netlists Used For Simulation

In the following, the circuit netlists are given. Each of the netlist contains only the circuit elements such as transistors and capacitors and the commands used in the simulations, such as "plot", "Fourier", and "AC", are not shown. The transistor model parameters are found in Appendix A, and they are not included in the netlists.

B.1 An N-Channel Differential Pair

Circuit shown in Figure 2.3

```
$***Input Differential  Pair**************
Bsim M1 1 2 3 3 MODEL NM W=30U L=8U PUBLIC
Bsim M2 1 2 3 3 MODEL NM W=30U L=8U PUBLIC

$***Bias Current**************************
Bsim M3 3 4 0 0 MODEL NM W=60U L=5U PUBLIC
Volt Vb  4 0 DC=1.3

$***Power Supply**************************
Volt VDD 1 0 DC=3.0 I=I_VDD

$***Input Common Mode Voltage*************
Volt vin 2 0 DC=VCM
```

B.2 A CMOS Source Follower

Circuit shown in Figure 2.5

```
$***Output Transistor**********************
Bsim M1 1 2 3 0 MODEL NM W=250U L=2U PUBLIC

$***Bias Current***************************
Bsim M2 3 4 0 0 MODEL NM W=20U L=2U PUBLIC
Volt Vb  4 0 DC=1.3

$***Power Supply***************************
Volt VDD 1 0 DC=3.0 I=I_VDD

$***Input  Voltage*************************
Volt vin 2 0 DC=VIN
```

B.3 A CMOS Rail-to-Rail Differential Pair

Circuit shown in Figure 2.7

```
$***P-Channel Input Differential  Pair*****
Bsim M1  0 2 5 1 MODEL P1 W=80U L=8U PUBLIC
Bsim M1a 0 2 5 1 MODEL P1 W=80U L=8U PUBLIC

$***N-Channel Input Differential  Pair*****
Bsim M2  1 2 3 0 MODEL NM W=30U L=8U PUBLIC
Bsim M2a 1 2 3 0 MODEL NM W=30U L=8U PUBLIC

$***Bias Currents *************************
Bsim M3 3 4 0 0 MODEL NM W=30U L=5U PUBLIC
Volt Vb1  4 0 DC=1.45

Bsim M4 5 6 1 0 MODEL P1 W=60U L=5U PUBLIC
Volt Vb2  6 0 DC=1.7

$***Power Supply***************************
Volt VDD 1 0 DC=3.0 I=I_VDD

$***Input Common Mode Voltage**************
Volt vin 2 0 DC=VCM
```

B.4 A Simple Operational Amplifier Model

Circuit shown in Figure 2.1

```
$***The First stage************
Res Rin 1 2 1000000000000
Csource gmT 4 0 1 2  .000025 VCCS
Res Ro1 4 0 10000000
Cap Co1 4 0 .5P

$***The Second Stage***********
Csource gmo 3 0 4 0  .00025 VCCS
Res Ro2 3 0 100000
Cap Co2 3 0 4P

$***Compensation Cap***********
Cap Cc 3 4 2P

$***Input  Voltage*************
Volt vin 1 0 AC=1
Volt vfed 2 0 DC=0
```

B.5 A Simple Rail-to-Rail Operational Amplifier With an Ideal Gain Stage

Circuit shown in Figure 2.9

```
$***P-Channel Input Differential  Pair********
Bsim M1   10 2 5 1  MODEL P1 W=80U L=8U PUBLIC
Bsim M1a  11 22 5 1 MODEL P1 W=80U L=8U PUBLIC

$***N-Channel Input Differential  Pair********
Bsim M2  12 2 3 0  MODEL NM W=30U L=8U PUBLIC
Bsim M2a 13 22 3 0 MODEL NM W=30U L=8U PUBLIC

***(These resistors are included to monitor
*** the currents of the input stage.
*** The currents will be used in
*** the gain stage)***************************
Res R1 10 0 .00001
Res R2 11 0 .00001
Res R3 1 12 .00001
```

```
Res R4 1 13 .00001

$***Bias Currents ****************************
Bsim M3 3 4 0 0 MODEL NM W=30U L=5U PUBLIC
Volt Vb1  4 0 DC=1.45

Bsim M4 5 6 1 0 MODEL P1 W=60U L=5U PUBLIC
Volt Vb2  6 0 DC=1.7

$***The Ideal Gain   Stage*******************
Csource Gm1 7 14 10 0 1 CCCS
Csource Gm2 7 14 11 0 -1 CCCS
Csource Gm3 7 14 1 12 -1 CCCS
Csource Gm4 7 14 1 13 1 CCCS
Volt V14 14 0 DC=1.5
Res Ro 7 0 3000000
Cap Co 7 0 7P

$***Power Supply*****************************
Volt VDD 1 0 DC=3.0 I=I_VDD

$***Input  Voltage***************************
Volt vin 2 0 DC=1.5 SS 1 .1
Volt vf  22 7 DC=0
```

B.6 The Second Constant-g_m Input Stage Using Square-Root Circuit

Circuit shown in Figure 3.3

```
$***Constant gm Biasing Circuit*****************
Bsim M3 13 13 8 100 MODEL NM W=20U L=2U
Bsim M4 14 13 7 100 MODEL NM W=20U L=2U
Bsim M5 12 12 7 100 MODEL NM W=20U L=2U
Bsim M6 16 12 8 100 MODEL NM W=20U L=2U
Bsim M7 7 17 9 101  MODEL P1 W=50U L=2U
Bsim M8 8 4 9 101   MODEL P1 W=50U L=2U

***Current Mirrors***
Bsim M9 14 14 101 101  MODEL P1 W=250U L=12U
Bsim M10 3 14 101 101  MODEL P1 W=250U L=12U

Bsim M11 15 15 100 100 MODEL NM W=150U L=5U
```

```
Bsim M12 4 15 100 100  MODEL NM W=150U L=5U

Bsim M13 16 16 101 101 MODEL P1 W=250U L=12U
Bsim M14 15 16 101 101 MODEL P1 W=250U L=12U

Bsim M15 7 7 100 100   MODEL NM W=150U L=5U
Bsim M16 8 7 100 100   MODEL NM W=150U L=5U

$***Constant Bias Currents and Voltages*********
Curr Iref1 101 13 DC=10U
Curr Iref2 101 12 DC=10U
Curr Iref3 101  9 DC=40U
Volt Vb 17 0      DC=.49

$***Power Supply Voltages**********************
Volt VDD 101 0 DC=3
Volt VSS 100 0 DC=0

$***Input Voltages******************************
Volt Vin1  1 0 DC=VCM
Volt Vin2  2 0 DC=VCM
```

B.7 Monitor Circuit 1

Circuit shown in Figure 4.7

```
$***P-Channel Input Differential  Pair******
Bsim M1  0 2 5 1 MODEL P1 W=80U L=8U PUBLIC
Bsim M1a 0 2 5 1 MODEL P1 W=80U L=8U PUBLIC

$***N-Channel Input Differential  Pair******
$Bsim M2  1 2 3 0 MODEL NM W=30U L=8U PUBLIC
$Bsim M2a 1 2 3 0 MODEL NM W=30U L=8U PUBLIC

$***Monitor Circuit 1**********************
$
$***Bias Current Iref***
Curr Iref 4 0 DC=20U
Bsim Mr 4 4 1 0 MODEL P1 W=60U L=15U PUBLIC
Bsim Mp 5 4 1 0 MODEL P1 W=60U L=15U PUBLIC
Bsim Mq 6 4 1 0 MODEL P1 W=60U L=15U PUBLIC
$
$***Replica of p-channel input diffy pair**
```

```
Bsim M3  0 2 6 1 MODEL P1 W=80U L=8U PUBLIC
Bsim M3a 0 2 6 1 MODEL P1 W=80U L=8U PUBLIC

$***Power Supply**************************
Volt VDD 1 0 DC=3.0 I=I_VDD

$***Input Common Mode Voltage*************
Volt vin 2 0 DC=VCM
```

B.8 Monitor Circuit 2

Circuit shown in Figure 4.8

```
$***P-Channel Input Differential  Pair******
Bsim M1  0 2 5 1 MODEL P1 W=80U L=8U PUBLIC
Bsim M1a 0 2 5 1 MODEL P1 W=80U L=8U PUBLIC

$***N-Channel Input Differential  Pair******
$Bsim M2  1 2 3 0 MODEL NM W=30U L=8U PUBLIC
$Bsim M2a 1 2 3 0 MODEL NM W=30U L=8U PUBLIC

$***Monitor Circuit 1**********************

Curr Iref 4 0 DC=20U
Volt Vb   6 0 DC=1.5
Bsim M3  1 2 4 0 MODEL NM W=90U L=5U PUBLIC
Bsim M3a 1 2 4 0 MODEL NM W=90U L=5U PUBLIC
Bsim Mb  7 6 4 0 MODEL NM W=60U L=5U PUBLIC

Bsim Mr 7 7 1 0 MODEL P1 W=60U L=15U PUBLIC
Bsim Mp 5 7 1 0 MODEL P1 W=60U L=15U PUBLIC
Bsim Mq 0 7 1 0 MODEL P1 W=60U L=15U PUBLIC

$***Power Supply**************************
Volt VDD 1 0 DC=3.0 I=I_VDD

$***Input Common Mode Voltage*************
Volt vin 2 0 DC=VCM
```

B.9 Monitor 1 With Constant-g_m Bias 2

Circuit shown in Figure 5.1

```
$***P-Channel Input Differential  Pair******
Bsim M1  0 2 5 1 MODEL P1 W=80U L=8U PUBLIC
Bsim M1a 0 2 5 1 MODEL P1 W=80U L=8U PUBLIC

$***N-Channel Input Differential  Pair******
Bsim M2  1 2 3 0 MODEL NM W=30U L=8U PUBLIC
Bsim M2a 1 2 3 0 MODEL NM W=30U L=8U PUBLIC

$***Monitor Circuit 1***********************
$
$***Bias Current Iref***
Curr Iref 4 0 DC=12U
Bsim Mr 4 4 1 0 MODEL P1 W=60U L=15U PUBLIC
Bsim Mp 5 4 1 0 MODEL P1 W=60U L=15U PUBLIC
Bsim Mq 6 4 1 0 MODEL P1 W=60U L=15U PUBLIC
$
$***Replica of p-channel input diffy pair**
Bsim M3b  7 2 6 1 MODEL P1 W=80U L=8U PUBLIC
Bsim M3a  7 2 6 1 MODEL P1 W=80U L=8U PUBLIC

$***Constant-gn Circuit 2******************
$
Bsim M3 7 7 0 0     MODEL NM W=160U L=8U PUBLIC
Bsim M4 8 7 9 1     MODEL P1 W=60U  L=8U
Bsim M5 0 10 9 1    MODEL P1 W=60U  L=8U
Bsim M6 10 10 0 0   MODEL NM W=160U L=8U
Bsim M7 8 8 0 0     MODEL NM W=100U L=5U
Bsim M8 3 8 0 0     MODEL NM W=100U L=5U
Bsim M9 11 8 0 0    MODEL NM W=100U L=5U
Bsim M10 11 11 1 1  MODEL P1 W=150U  L=10U
Bsim M11 9 11 1 1   MODEL P1 W=150U  L=10U
$
Curr Ic 1 10 DC=3U
Curr Id 1 9  DC=3U

$***Power Supply**************************
Volt VDD 1 0 DC=3.0 I=I_VDD

$***Input Common Mode Voltage*************
Volt vin 2 0 DC=VCM
```

B.10 Constant-g_m Input Stage 1

Circuit shown in Figure 5.9

```
$***Input Differential Pairs**********************
Bsim M1  0 2 4 1   MODEL P1 W=75U L=8U PUBLIC
Bsim M1a 0 109 4 1 MODEL P1 W=75U L=8U
Bsim M2  1 2 5 0   MODEL NM W=30U L=8U PUBLIC
Bsim M2a 1 109 5 0 MODEL NM W=30U L=8U

$***Monitor Circuit 1*****************************
Bsim Mp   4 8 1 1   MODEL P1 W=400U L=4U
Bsim Mq   44 8 1 1  MODEL P1 W=200U L=4U
Bsim M3a  11 2 44 1 MODEL P1 W=38U L=8U
Bsim M3b  11 2 44 1 MODEL P1 W=38U L=8U

$***Bias Circuit 2********************************
Bsim M3 11 11 0 0   MODEL NM W=30U L=8U  PUBLIC
Bsim M4 13 11 12 1  MODEL P1 W=75U L=8U PUBLIC
Bsim M5 0 21 84 1   MODEL P1 W=75U L=8U PUBLIC
Bsim M7 13 13 0 0   MODEL NM W=30U L=5U PUBLIC
Bsim M8 5 13 0 0    MODEL NM W=60U L=5U PUBLIC
Bsim M9 84 13 0 0   MODEL NM W=30U L=5U
Bsim M10 16 16 84 1 MODEL P1 W=100U L=4U
Bsim M11 17 16 12 1 MODEL P1 W=100U L=4U
Bsim M12 16 17 0 0  MODEL NM W=40U L=12U
Bsim M13 17 17 0 0  MODEL NM W=40U L=12U
Curr IBIAS2 1 84 DC=25U
Curr IBIAS3 1 12 DC=20U

$***Inmax--Ipmax Circuit**************************
Bsim MB1  21 21 0 0  MODEL NM W=30.0U L=8U PUBLIC
Bsim MB2  80 27 0 0  MODEL NM W=30U L=8U PUBLIC
Bsim MB3  27 27 20 1 MODEL P1 W=75U L=8U
Bsim MB4  26 21 20 1 MODEL P1 W=75U L=8U
Bsim MB5  27 26 0 0  MODEL NM W=25U L=20U  PUBLIC
Bsim MB6  26 26 0 0  MODEL NM W=100U L=20U PUBLIC
Bsim MB7  81 8 1 1   MODEL P1 W=50U L=4U
Bsim MB8  8 8 1 1    MODEL P1 W=200U L=4U
Bsim MB9  21 80 81 1 MODEL P1 W=50U L=2U
Bsim MB10 80 80 8 1  MODEL P1 W=200U L=2U
Curr IBIAS1 1 20 DC=25U

$***Power Supply**********************************
Volt VDD 1 0 DC=3.0 I=I_VDD
```

```
$***Input Voltages********************************
Volt Vin1  2   0 DC VCM
Volt Vin2  109 2 DC 0
```

B.11 Constant-g_m Input Stage 2

Circuit shown in Figure 5.11

```
$***Input Differential Pairs**********************
Bsim M1  0 2 4 1   MODEL P1 W=75U L=8U PUBLIC
Bsim M1a 0 109 4 1 MODEL P1 W=75U L=8U
Bsim M2  1 2 5 0   MODEL NM W=30U L=8U PUBLIC
Bsim M2a 1 109 5 0 MODEL NM W=30U L=8U

$***Monitor Circuit 2*****************************
Bsim Mp   4 7 1 1  MODEL P1 W=200U L=5U
Bsim Mq   11 7 1 1 MODEL P1 W=100U L=5U
Bsim Mr   7 7 1 1  MODEL P1 W=100U L=5U
Bsim M3a  1 2 3 0  MODEL NM W=45U L=4U
Bsim M3b  1 2 3 0  MODEL NM W=45U L=4U
Bsim Mb   7 6 3 0  MODEL NM W=30U L=4U
Bsim MB9  3 21 0 0 MODEL NM W=120U L=8U
Volt Vb 6 0 DC=1.7 I=I_Vb

$***Bias Circuit 2********************************
Bsim M3   11 11 0 0  MODEL NM W=30U L=8U
Bsim M4   13 11 12 1 MODEL P1 W=75U L=8U
Bsim M5   0 21 84 1  MODEL P1 W=75U L=8U
Bsim M7   13 13 0 0  MODEL NM W=40U L=15U
Bsim M8   5 13 0 0   MODEL NM W=80U L=15U
Bsim M9   84 13 0 0  MODEL NM W=40U L=15U
Bsim M10  16 16 84 1 MODEL P1 W=50U L=4U
Bsim M11  17 16 12 1 MODEL P1 W=50U L=4U
Bsim M12  16 17 0 0  MODEL NM W=20U L=5U
Bsim M13  17 17 0 0  MODEL NM W=20U L=5U
Curr IBIAS2 1 84 DC=25U
Curr IBIAS3 1 12 DC=20U

$***Inmax--Ipmax Circuit**************************
Bsim MB1  21 21 0 0  MODEL NM W=30.0U L=8U PUBLIC
Bsim MB2  80 27 0 0  MODEL NM W=30U L=8U PUBLIC
Bsim MB3  27 27 20 1 MODEL P1 W=75U L=8U
Bsim MB4  26 21 20 1 MODEL P1 W=75U L=8U
Bsim MB5  27 26 0 0  MODEL NM W=25U L=20U  PUBLIC
```

```
Bsim MB6   26 26 0 0  MODEL NM W=100U L=20U PUBLIC
Bsim MB7   81 8 1 1   MODEL P1 W=50U L=10U
Bsim MB8   8 8 1 1    MODEL P1 W=200U L=10U
Bsim MB10   21 80 81  1 MODEL P1 W=50U L=2U
Bsim MB11 80 80 8 1  MODEL P1 W=200U L=2U
Curr IBIAS1 1 20 DC=25U

$***Power Supply***********************************
Volt VDD 1 0 DC=3.0 I=I_VDD

$***Input Voltages*********************************
Volt Vin1  2   0 DC VCM
Volt Vin2  109 2 DC 0
```

B.12 Small Signal Model of the Modified Output Stage

Circuit shown in Figure 6.6

```
***Dependent Current Source for gmT Vin/2****
Csource G1 1 0 10 0 .00005 VCCS

***Dependent Current Source for -gmT Vin/2***
Csource G2 4 0 10 0 -.00005 VCCS

***Dependent Current Source for gmc3 Vc******
Csource G3 4 6 4 0  .0002 VCCS

***Dependent Current Source for gmc4 Vb******
Csource G4 1 3 1 0 .0002 VCCS

***Dependent Current Source for gmo6 V1-*****
Csource G5 5 0 6 0 .0002 VCCS

***Dependent Current Source for gm02 V2******
Csource G6 2 0 5 0 .0002 VCCS

***Dependent Current Source for gmo1 V1+*****
Csource G7 2 0 3 0 .0002 VCCS

***Resistors*********************************
Res Rin      10 0 1000000000
```

```
Res roc1     4 0 300000
Res roc2     1 0 300000
Res roc3     4 6 300000
Res roc4     1 3 300000
Res 2Ro1     6 0 40000000
Res 2Ro11    3 0 40000000
Res Ro       2 0 10000
Res 1/gmo5   5 0 5000

***Capacitors********************************
Cap Cc1   1 2 2P
Cap Cc2   4 5 2P
Cap Cp    5 0 .1P
Cap Co1   6 0 .1P
Cap Co11  3 0 .1P
Cap Cl    2 0 10P

Volt Vin 10 0 DC=0 AC=1
```

B.13 Modified Class AB Controlled Output Stage

Circuit shown in Figure 6.9

```
***Output Stage**************************************
*
***Common Source Amplifier*********************
Bsim M01 72 208 1 1 MODEL P1 W=100U L=2U PUBLIC
Bsim M02 72 79 0 0  MODEL NM W=40U  L=2U PUBLIC
*
***Class AB Control*******************************
Bsim M03  1 76 75 0     MODEL NM W=100U L=4U PUBLIC
Bsim M04  202 171 75 0 MODEL NM W=100U L=4U PUBLIC
Bsim M04B 203 171 75 0 MODEL NM W=100U L=4U PUBLIC
Bsim M05  79 79 0 0     MODEL NM W=20U  L=2U PUBLIC
Bsim M06  79 217 1 1    MODEL P1 W=50U  L=2U PUBLIC
Bsim M07  77 78 0 0     MODEL NM W=40U  L=4U PUBLIC
Bsim M07B 78 78 0 0     MODEL NM W=40U  L=4U PUBLIC
Bsim M08  76 208 1 1    MODEL P1 W=50U  L=2U PUBLIC
Bsim M09  73 79 0 0     MODEL NM W=20U  L=2U PUBLIC
Bsim M010 77 79 0 0     MODEL NM W=20U  L=2U PUBLIC
```

```
Bsim MO11 73 73 74 1   MODEL P1 W=50U  L=4U PUBLIC
Bsim MO12 0 73 171 1   MODEL P1 W=50U  L=4U PUBLIC
Bsim MO13 78 77 76 1   MODEL P1 W=50U  L=4U PUBLIC
Bsim MO14 77 77 74 1   MODEL P1 W=50U  L=4U PUBLIC
*
Volt Vc5 74 0 DC=2.5
Curr Imin  1 171 DC 5U
Curr IB 75 0 DC=30U
***IB1 and IB2 can be supplied by MC1 and MC2,
***but they are separately placed in this file.
Curr IB1 1 202  DC=10U
Curr IB2 1 203  DC=10U

***Cascode****************************************
Bsim MC1 202 204 1 1   MODEL P1 W=100U L=8U PUBLIC
Bsim MC2 203 204 1 1   MODEL P1 W=100U L=8U PUBLIC
Bsim MC3 217 205 202 1 MODEL P1 W=550U L=4U PUBLIC
Bsim MC4 208 205 203 1 MODEL P1 W=550U L=4U PUBLIC
Bsim MC5 217 207 200 0 MODEL NM W=200U L=4U PUBLIC
Bsim MC6 208 207 201 0 MODEL NM W=200U L=4U PUBLIC
Bsim MC7 200 206 0 0   MODEL NM W=100U L=8U PUBLIC
Bsim MC8 201 206 0 0   MODEL NM W=100U L=8U PUBLIC
*
Volt VC1 204 0 DC 1.55
Volt VC2 205 0 DC 1.3
Volt VC3 207 0 DC 1.8
Volt VC4 206 0 DC 1.35

***Ideal Input Stage************
Csource  gmn/2  203 0 2 22 .000025  VCCS
Csource  gmnn/2 202 0 2 22 -.000025 VCCS
Csource  gmp/2  200 0 2 22 -.000025 VCCS
Csource  gmpp/2 201 0 2 22 .000025  VCCS
Res Rin 2 22 1000000000000
Volt Vref 500 0  DC 1.5
Volt Vin  2 0  DC=VCM
Volt Vfed 22 72  DC=0

***Power Supply***********************************
Volt VDD 1 0 DC=3 I=I_VDD

***Load**************
Res Rout 72 500 10000
```

B.14 Opamp 1

Circuit shown in Figure 7.1

```
***Input Differential Pairs*************************
Bsim M1  200 2 4 1   MODEL P1 W=50U L=8U PUBLIC
Bsim M1a 201 22 4 1  MODEL P1 W=50U L=8U PUBLIC
Bsim M2  202 2 5 60  MODEL NM W=20U L=8U PUBLIC
Bsim M2a 203 22 5 60 MODEL NM W=20U L=8U PUBLIC

**Bias Currents************************************
Bsim Mr  7 7 1 1 MODEL P1 W=100U L=5U
Bsim Mp  4 7 1 1 MODEL P1 W=200U L=5U
Curr Ip  7 0 DC=15U

Curr In  1 13 DC=15U
Bsim M7  13 13 60 60 MODEL NM W=40U L=15U
Bsim M8  5 13 60 60  MODEL NM W=80U L=15U

***Cascode Gain Stage******************************
Bsim MC1 202 204 1 1    MODEL P1 W=100U L=8U PUBLIC
Bsim MC2 203 204 1 1    MODEL P1 W=100U L=8U PUBLIC
Bsim MC3 217 205 202 1  MODEL P1 W=500U L=4U PUBLIC
Bsim MC4 208 205 203  1 MODEL P1 W=500U L=4U PUBLIC
Bsim MC5 217 207 200 60 MODEL NM W=200U L=4U PUBLIC
Bsim MC6 208 207 201 60 MODEL NM W=200U L=4U PUBLIC
Bsim MC7 200 206 60 60  MODEL NM W=100U L=8U PUBLIC
Bsim MC8 201 206 60 60  MODEL NM W=100U L=8U PUBLIC
*
Volt VC1 204 0 DC 1.521234
Volt VC2 205 0 DC 1.3
Volt VC3 207 0 DC 1.8
Volt V06 217 206 DC 0

***Load*******************************************
Cap Cl1 208 70 5P

Volt Vref 70 0 DC=1.5

***Voltage Sources*********************************
Volt VDD 1 0  DC=3.
Volt VSS 60 0  DC=0.
```

B.15 Opamp 1a

Circuit shown in Figure 7.6

```
*********Constant-gm Input Stage******************
*
$***Input Differential Pairs**********************
Bsim M1  200 2 4 1    MODEL P1 W=75U L=8U PUBLIC
Bsim M1a 201 22 4 1   MODEL P1 W=75U L=8U PUBLIC
Bsim M2  202 2 5 60   MODEL NM W=30U L=8U PUBLIC
Bsim M2a 203 22 5 60  MODEL NM W=30U L=8U PUBLIC

$***Monitor Circuit 1*****************************
Bsim Mp   4 8 1 1    MODEL P1 W=400U L=4U
Bsim Mq   44 8 1 1   MODEL P1 W=200U L=4U
Bsim M3a  11 2 44 1  MODEL P1 W=38U L=8U
Bsim M3b  11 22 44 1 MODEL P1 W=38U L=8U

$***Bias Circuit 2********************************
Bsim M3 11 11 60 60   MODEL NM W=30U L=8U  PUBLIC
Bsim M4 13 11 12 1    MODEL P1 W=75U L=8U PUBLIC
Bsim M5 60 21 84 1    MODEL P1 W=75U L=8U PUBLIC
Bsim M7 13 13 60 60   MODEL NM W=30U L=5U PUBLIC
Bsim M8 5 13 60 60    MODEL NM W=60U L=5U PUBLIC
Bsim M9 84 13 60 60   MODEL NM W=30U L=5U
Bsim M10 16 16 84 1   MODEL P1 W=100U L=4U
Bsim M11 17 16 12 1   MODEL P1 W=100U L=4U
Bsim M12 16 17 60 60  MODEL NM W=40U L=12U
Bsim M13 17 17 60 60  MODEL NM W=40U L=12U
Curr IBIAS2 1 84 DC=25U
Curr IBIAS3 1 12 DC=20U

$***Inmax--Ipmax Circuit**************************
Bsim MB1  21 21 60 60  MODEL NM W=30.0U L=8U PUBLIC
Bsim MB2  80 27 60 60  MODEL NM W=30U L=8U PUBLIC
Bsim MB3  27 27 20 1   MODEL P1 W=75U L=8U
Bsim MB4  26 21 20 1   MODEL P1 W=75U L=8U
Bsim MB5  27 26 60 60  MODEL NM W=25U L=20U  PUBLIC
Bsim MB6  26 26 60 60  MODEL NM W=100U L=20U PUBLIC
Bsim MB7  81 8 1 1     MODEL P1 W=50U L=4U
Bsim MB8  8 8 1 1      MODEL P1 W=200U L=4U
Bsim MB9  21 80 81 1   MODEL P1 W=50U L=2U
Bsim MB10 80 80 8 1    MODEL P1 W=200U L=2U
Curr IBIAS1 1 20 DC=25U

***Cascode Gain Stage*****************************
```

```
Bsim MC1 202 204 1 1    MODEL P1 W=100U L=8U PUBLIC
Bsim MC2 203 204 1 1    MODEL P1 W=100U L=8U PUBLIC
Bsim MC3 217 205 202 1  MODEL P1 W=500U L=4U PUBLIC
Bsim MC4 208 205 203  1 MODEL P1 W=500U L=4U PUBLIC
Bsim MC5 217 207 200 60 MODEL NM W=200U L=4U PUBLIC
Bsim MC6 208 207 201 60 MODEL NM W=200U L=4U PUBLIC
Bsim MC7 200 206 60 60  MODEL NM W=100U L=8U PUBLIC
Bsim MC8 201 206 60 60  MODEL NM W=100U L=8U PUBLIC
Volt VC1 204 0 DC 1.521234
Volt VC2 205 0 DC 1.3
Volt VC3 207 0 DC 1.8
Volt V06 217 206 DC 0

***Load******************
Cap Cl1 208 50 5P


***Voltage Sources********
Volt VDD 1 0  DC=3.
Volt VSS 60 0 DC=0 AC=1
```

B.16 Opamp 1b

Circuit shown in Figure 7.9

```
***Constant-gm Input Stage 2**********************
*
$***Input Differential Pairs**********************
Bsim M1  200 2 4 1    MODEL P1 W=75U L=8U PUBLIC
Bsim M1a 201 22 4 1   MODEL P1 W=75U L=8U PUBLIC
Bsim M2  202 2 5 60   MODEL NM W=30U L=8U PUBLIC
Bsim M2a 203 22 5 60  MODEL NM W=30U L=8U PUBLIC
*
$***Monitor Circuit 2*****************************
Bsim Mp   4 7 1 1    MODEL P1 W=200U L=5U
Bsim Mq   11 7 1 1   MODEL P1 W=100U L=5U
Bsim Mr   7 7 1 1    MODEL P1 W=100U L=5U
Bsim M3a  1 2 3 60   MODEL NM W=45U L=4U
Bsim M3b  1 22 3 60  MODEL NM W=45U L=4U
Bsim Mb   7 6 3 60   MODEL NM W=30U L=4U
Bsim MB9  3 21 60 60 MODEL NM W=120U L=8U
Volt Vb 6 0 DC=1.6 I=I_Vb
```

```
*
$***Bias Circuit 2********************************
Bsim M3   11 11 60 60  MODEL NM W=30U L=8U
Bsim M4   13 11 12 1   MODEL P1 W=75U L=8U
Bsim M5   60 21 84 1   MODEL P1 W=75U L=8U
Bsim M7   13 13 60 60  MODEL NM W=40U L=15U
Bsim M8   5 13 60 60   MODEL NM W=80U L=15U
Bsim M9   84 13 60 60  MODEL NM W=40U L=15U
Bsim M10  16 16 84 1   MODEL P1 W=50U L=4U
Bsim M11  17 16 12 1   MODEL P1 W=50U L=4U
Bsim M12  16 17 60 60  MODEL NM W=20U L=5U
Bsim M13  17 17 60 60  MODEL NM W=20U L=5U
Curr IBIAS2 1 84 DC=25U
Curr IBIAS3 1 12 DC=20U
*
$***Inmax--Ipmax Circuit*************************
Bsim MB1  21 21 60 60   MODEL NM W=30.0U L=8U PUBLIC
Bsim MB2  80 27 60 60   MODEL NM W=30U L=8U PUBLIC
Bsim MB3  27 27 20 1    MODEL P1 W=75U L=8U
Bsim MB4  26 21 20 1    MODEL P1 W=75U L=8U
Bsim MB5  27 26 60 60   MODEL NM W=25U L=20U  PUBLIC
Bsim MB6  26 26 60 60   MODEL NM W=100U L=20U PUBLIC
Bsim MB7  81 8 1 1      MODEL P1 W=50U L=10U
Bsim MB8  8 8 1 1       MODEL P1 W=200U L=10U
Bsim MB10  21 80 81 1   MODEL P1 W=50U L=2U
Bsim MB11 80 80 8 1     MODEL P1 W=200U L=2U
Curr IBIAS1 1 20 DC=25U

***Cascode Gain Stage******************************
Bsim MC1 202 204 1 1     MODEL P1 W=100U L=8U PUBLIC
Bsim MC2 203 204 1 1     MODEL P1 W=100U L=8U PUBLIC
Bsim MC3 217 205 202 1   MODEL P1 W=500U L=4U PUBLIC
Bsim MC4 208 205 203  1  MODEL P1 W=500U L=4U PUBLIC
Bsim MC5 217 207 200 60  MODEL NM W=200U L=4U PUBLIC
Bsim MC6 208 207 201 60  MODEL NM W=200U L=4U PUBLIC
Bsim MC7 200 206 60 60   MODEL NM W=100U L=8U PUBLIC
Bsim MC8 201 206 60 60   MODEL NM W=100U L=8U PUBLIC
Volt VC1 204 0 DC 1.521234
Volt VC2 205 0 DC 1.3
Volt VC3 207 0 DC 1.8
Volt V06 217 206 DC 0

***Load*********
Cap Cl1 208 50 5P
```

```
***Power Supply********
Volt VDD  1 0   DC=3.
Volt VSS  60 0 DC=0
```

B.17 Opamp 2

Circuit shown in Figure 8.1

```
***Output Stage*****************************************
*
***Common Source Amplifier*************************
Bsim M01 72 208 1 1   MODEL P1 W=100U L=2U PUBLIC
Bsim M02 72 79 60 60  MODEL NM W=40U  L=2U PUBLIC
*
***Class AB Control*********************************
Bsim M03  1 76 75 60     MODEL NM W=100U L=4U PUBLIC
Bsim M04  202 171 75 60  MODEL NM W=100U L=4U PUBLIC
Bsim M04B 203 171 75 60  MODEL NM W=100U L=4U PUBLIC
Bsim M05  79 79 60 60    MODEL NM W=20U  L=2U PUBLIC
Bsim M06  79 217 1 1     MODEL P1 W=50U  L=2U PUBLIC
Bsim M07  77 78 60 60    MODEL NM W=40U  L=4U PUBLIC
Bsim M07B 78 78 60 60    MODEL NM W=40U  L=4U PUBLIC
Bsim M08  76 208 1 1     MODEL P1 W=50U  L=2U PUBLIC
Bsim M09  73 79 60 60    MODEL NM W=20U  L=2U PUBLIC
Bsim M010 77 79 60 60    MODEL NM W=20U  L=2U PUBLIC
Bsim M011 73 73 74 1     MODEL P1 W=50U  L=4U PUBLIC
Bsim M012 60 73 171 1    MODEL P1 W=50U  L=4U PUBLIC
Bsim M013 78 77 76 1     MODEL P1 W=50U  L=4U PUBLIC
Bsim M014 77 77 74 1     MODEL P1 W=50U  L=4U PUBLIC
*
Volt Vc5 74 0 DC=2.5
Curr Imin  1 171 DC 5U
Curr IB 75 0 DC=30U
***IB1 and IB2 can be supplied by MC1 and MC2,
***but they are separately placed in this file.
Curr IB1 1 202  DC=10U
Curr IB2 1 203  DC=10U
***Cascode Gain Stage*****************************
Bsim MC1 202 204  1 1   MODEL P1 W=100U L=8U PUBLIC
Bsim MC2 203 204 1 1    MODEL P1 W=100U L=8U PUBLIC
Bsim MC3 217 205 202 1  MODEL P1 W=550U L=4U PUBLIC
```

```
Bsim MC4 208 205 203 1  MODEL P1 W=550U L=4U PUBLIC
Bsim MC5 217 207 200 60 MODEL NM W=200U L=4U PUBLIC
Bsim MC6 208 207 201 60 MODEL NM W=200U L=4U PUBLIC
Bsim MC7 200 206 60 60  MODEL NM W=100U L=8U PUBLIC
Bsim MC8 201 206 60 60  MODEL NM W=100U L=8U PUBLIC

Volt VC1 204 0 DC 1.55
Volt VC2 205 0 DC 1.3
Volt VC3 207 0 DC 1.8
Volt V06 206 0 DC 1.35


***Input Differential Pairs*************************
Bsim M1  200 2 4 1  MODEL P1 W=50U L=8U PUBLIC
Bsim M1a 201 22 4 1 MODEL P1 W=50U L=8U PUBLIC
Bsim M1b 120 2 4 1  MODEL P1 W=25U L=16U PUBLIC
Bsim M1c 120 22 4 1 MODEL P1 W=25U L=16U PUBLIC

Bsim M2 202 2 5 60   MODEL NM W=20U L=8U PUBLIC
Bsim M2a 203 22 5 60 MODEL NM W=20U L=8U PUBLIC
Bsim M2b 121 2 5 60  MODEL NM W=10U L=16U PUBLIC
Bsim M2c 121 22 5 60 MODEL NM W=10U L=16U PUBLIC

Bsim ME1 202 121 1 1 MODEL P1 W=200U L=4U PUBLIC
Bsim ME2 121 121 1 1 MODEL P1 W=100U L=4U PUBLIC
Bsim ME3 203 121 1 1 MODEL P1 W=200U L=4U PUBLIC

Bsim ME4 200 120 60 60 MODEL NM W=52U L=4U PUBLIC
Bsim ME5 120 120 60 60 MODEL NM W=26U L=4U PUBLIC
Bsim ME6 201 120 60 60 MODEL NM W=52U L=4U PUBLIC
*
Bsim M4 7 7 1 1 MODEL P1 W=100U L=5U
Bsim M7 4 7 1 1 MODEL P1 W=200U L=5U
Curr Ip 7 0 DC=15U

Curr In 1 13 DC=15U
Bsim M12 13 13 0 0 MODEL NM W=40U L=15U
Bsim M14 5 13 0 0  MODEL NM W=80U L=15U

***Load*******************************************
Cap Cl1   72  500 30P
Res Rout  72 500 100000
Volt Vref 500 0 DC 1.5

***Compensation Caps*****
```

```
Cap Cc1 72 201 2P
Cap Cc2 72 203 2P
Cap Cc3 79 202 2P
Cap Cc4 79 200 2P

***Power Supply**********
Volt VDD 1 0  DC=3
Volt VSS 60 0 DC=0
```

B.18 Opamp 2a

Circuit shown in Figure 8.4

```
***Output Stage*******************************************
*
***Common Source Amplifier*************************
Bsim MO1 72 208 1 1   MODEL P1 W=100U L=2U PUBLIC
Bsim MO2 72 79 60 60  MODEL NM W=40U  L=2U PUBLIC
*
***Class AB Control**********************************
Bsim MO3  1 76 75 60     MODEL NM W=100U L=4U PUBLIC
Bsim MO4  202 171 75 60  MODEL NM W=100U L=4U PUBLIC
Bsim MO4B 203 171 75 60  MODEL NM W=100U L=4U PUBLIC
Bsim MO5  79 79 60 60    MODEL NM W=20U  L=2U PUBLIC
Bsim MO6  79 217 1 1     MODEL P1 W=50U  L=2U PUBLIC
Bsim MO7  77 78 60 60    MODEL NM W=40U  L=4U PUBLIC
Bsim MO7B 78 78 60 60    MODEL NM W=40U  L=4U PUBLIC
Bsim MO8  76 208 1 1     MODEL P1 W=50U  L=2U PUBLIC
Bsim MO9  73 79 60 60    MODEL NM W=20U  L=2U PUBLIC
Bsim MO10 77 79 60 60    MODEL NM W=20U  L=2U PUBLIC
Bsim MO11 73 73 74 1     MODEL P1 W=50U  L=4U PUBLIC
Bsim MO12 60 73 171 1    MODEL P1 W=50U  L=4U PUBLIC
Bsim MO13 78 77 76 1     MODEL P1 W=50U  L=4U PUBLIC
Bsim MO14 77 77 74 1     MODEL P1 W=50U  L=4U PUBLIC
*
Volt Vc5 74 0 DC=2.5
Curr Imin  1 171 DC 5U
Curr IB 75 0 DC=30U
***IB1 and IB2 can be supplied by MC1 and MC2,
***but they are separately placed in this file.
Curr IB1 1 202  DC=10U
Curr IB2 1 203  DC=10U
***Cascode Gain Stage******************************
Bsim MC1 202 204  1 1   MODEL P1 W=100U L=8U PUBLIC
```

```
Bsim MC2 203 204 1 1    MODEL P1 W=100U L=8U PUBLIC
Bsim MC3 217 205 202 1  MODEL P1 W=550U L=4U PUBLIC
Bsim MC4 208 205 203 1  MODEL P1 W=550U L=4U PUBLIC
Bsim MC5 217 207 200 60 MODEL NM W=200U L=4U PUBLIC
Bsim MC6 208 207 201 60 MODEL NM W=200U L=4U PUBLIC
Bsim MC7 200 206 60 60  MODEL NM W=100U L=8U PUBLIC
Bsim MC8 201 206 60 60  MODEL NM W=100U L=8U PUBLIC

Volt VC1 204 0 DC 1.55
Volt VC2 205 0 DC 1.3
Volt VC3 207 0 DC 1.8
Volt V06 206 0 DC 1.35


*********Constant-gm Input Stage*****************
*
$***Input Differential Pairs*********************
Bsim M1  200 2 4 1    MODEL P1 W=75U L=8U PUBLIC
Bsim M1a 201 22 4 1   MODEL P1 W=75U L=8U PUBLIC
Bsim M2  202 2 5 60   MODEL NM W=30U L=8U PUBLIC
Bsim M2a 203 22 5 60  MODEL NM W=30U L=8U PUBLIC
*
***The circuit which keeps the Icascode constant**
Bsim ME1 202 209 1 1  MODEL P1 W=60U L=10U  PUBLIC
Bsim ME2 203 209 1 1  MODEL P1 W=60U L=10U PUBLIC
Bsim ME3 209 209 1 1  MODEL P1 W=60U L=10U PUBLIC
Bsim ME4 200 11 60 60 MODEL NM W=30U L=8U
Bsim ME5 201 11 60 60 MODEL NM W=30U L=8U
Bsim ME6 209 13 60 60 MODEL NM W=30U L=5U
*
$***Monitor Circuit 1****************************
Bsim Mp   4 8 1 1   MODEL P1 W=400U L=4U
Bsim Mq   44 8 1 1  MODEL P1 W=200U L=4U
Bsim M3a  11 2 44 1 MODEL P1 W=38U L=8U
Bsim M3b  11 22 44 1 MODEL P1 W=38U L=8U
*
$***Bias Circuit 2*******************************
Bsim M3 11 11 60 60   MODEL NM W=30U L=8U  PUBLIC
Bsim M4 13 11 12 1    MODEL P1 W=75U L=8U PUBLIC
Bsim M5 60 21 84 1    MODEL P1 W=75U L=8U PUBLIC
Bsim M7 13 13 60 60   MODEL NM W=30U L=5U PUBLIC
Bsim M8 5 13 60 60    MODEL NM W=60U L=5U PUBLIC
Bsim M9 84 13 60 60   MODEL NM W=30U L=5U
Bsim M10 16 16 84 1   MODEL P1 W=100U L=4U
Bsim M11 17 16 12 1   MODEL P1 W=100U L=4U
```

```
Bsim M12 16 17 60 60  MODEL NM W=40U L=12U
Bsim M13 17 17 60 60  MODEL NM W=40U L=12U
Curr IBIAS2 1 84 DC=25U
Curr IBIAS3 1 12 DC=20U
*
$***Inmax--Ipmax Circuit**************************
Bsim MB1  21 21 60 60  MODEL NM W=30.0U L=8U PUBLIC
Bsim MB2  80 27 60 60  MODEL NM W=30U L=8U PUBLIC
Bsim MB3  27 27 20 1   MODEL P1 W=75U L=8U
Bsim MB4  26 21 20 1   MODEL P1 W=75U L=8U
Bsim MB5  27 26 60 60  MODEL NM W=25U L=20U  PUBLIC
Bsim MB6  26 26 60 60  MODEL NM W=100U L=20U PUBLIC
Bsim MB7  81 8 1 1     MODEL P1 W=50U L=4U
Bsim MB8  8 8 1 1      MODEL P1 W=200U L=4U
Bsim MB9  21 80 81 1   MODEL P1 W=50U L=2U
Bsim MB10 80 80 8 1    MODEL P1 W=200U L=2U
Curr IBIAS1 1 20 DC=25U

***Load**********************************************
Cap Cl1   72  500 5P
Res Rout  72 500 100000
Volt Vref 500 0 DC 1.5

***Compensation Caps*****
Cap Cc1 72 201 2P
Cap Cc2 72 203 2P
Cap Cc3 79 202 2P
Cap Cc4 79 200 2P

Volt VDD 1 0  DC=3
Volt VSS 60 0 DC=0
```

B.19 Opamp 2b

Circuit shown in Figure 8.7

```
***Output Stage*****************************************
*
***Common Source Amplifier*************************
Bsim MO1 72 208 1 1   MODEL P1 W=100U L=2U PUBLIC
Bsim MO2 72 79 60 60  MODEL NM W=40U  L=2U PUBLIC
*
***Class AB Control***********************************
```

```
Bsim M03  1 76 75 60      MODEL NM W=100U L=4U PUBLIC
Bsim M04  202 171 75 60   MODEL NM W=100U L=4U PUBLIC
Bsim M04B 203 171 75 60   MODEL NM W=100U L=4U PUBLIC
Bsim M05  79 79 60 60     MODEL NM W=20U  L=2U PUBLIC
Bsim M06  79 217 1 1      MODEL P1 W=50U  L=2U PUBLIC
Bsim M07  77 78 60 60     MODEL NM W=40U  L=4U PUBLIC
Bsim M07B 78 78 60 60     MODEL NM W=40U  L=4U PUBLIC
Bsim M08  76 208 1 1      MODEL P1 W=50U  L=2U PUBLIC
Bsim M09  73 79 60 60     MODEL NM W=20U  L=2U PUBLIC
Bsim M010 77 79 60 60     MODEL NM W=20U  L=2U PUBLIC
Bsim M011 73 73 74 1      MODEL P1 W=50U  L=4U PUBLIC
Bsim M012 60 73 171 1     MODEL P1 W=50U  L=4U PUBLIC
Bsim M013 78 77 76 1      MODEL P1 W=50U  L=4U PUBLIC
Bsim M014 77 77 74 1      MODEL P1 W=50U  L=4U PUBLIC
*
Volt Vc5 74 0 DC=2.5
Curr Imin  1 171 DC 5U
Curr IB 75 0 DC=30U
***IB1 and IB2 can be supplied by MC1 and MC2,
***but they are separately placed in this file.
Curr IB1 1 202  DC=10U
Curr IB2 1 203  DC=10U

***Cascode Gain Stage******************************
Bsim MC1 202 204  1 1   MODEL P1 W=100U L=8U PUBLIC
Bsim MC2 203 204 1 1    MODEL P1 W=100U L=8U PUBLIC
Bsim MC3 217 205 202 1  MODEL P1 W=550U L=4U PUBLIC
Bsim MC4 208 205 203 1  MODEL P1 W=550U L=4U PUBLIC
Bsim MC5 217 207 200 60 MODEL NM W=200U L=4U PUBLIC
Bsim MC6 208 207 201 60 MODEL NM W=200U L=4U PUBLIC
Bsim MC7 200 206 60 60  MODEL NM W=100U L=8U PUBLIC
Bsim MC8 201 206 60 60  MODEL NM W=100U L=8U PUBLIC

Volt VC1 204 0 DC 1.55
Volt VC2 205 0 DC 1.3
Volt VC3 207 0 DC 1.8
Volt V06 206 0 DC 1.35

***Input Stage*************************************
*
$***Input Differential Pairs**********************
Bsim M1  200 2 4 1    MODEL P1 W=75U L=8U PUBLIC
Bsim M1a 201 22 4 1   MODEL P1 W=75U L=8U PUBLIC
Bsim M2  202 2 5 60   MODEL NM W=30U L=8U PUBLIC
Bsim M2a 203 22 5 60  MODEL NM W=30U L=8U PUBLIC
```

```
*
***The circuit which keeps the Icascode constant**
Bsim ME1 202 209 1 1  MODEL P1 W=60U L=10U  PUBLIC
Bsim ME2 203 209 1 1  MODEL P1 W=60U L=10U PUBLIC
Bsim ME3 209 209 1 1  MODEL P1 W=60U L=10U PUBLIC
Bsim ME4 200 11 60 60 MODEL NM W=30U L=8U
Bsim ME5 201 11 60 60 MODEL NM W=30U L=8U
Bsim ME6 209 13 60 60 MODEL NM W=30U L=5U
*
$***Monitor Circuit 2****************************
Bsim Mp   4 7 1 1   MODEL P1 W=200U L=5U
Bsim Mq   11 7 1 1  MODEL P1 W=100U L=5U
Bsim Mr   7 7 1 1   MODEL P1 W=100U L=5U
Bsim M3a  1 2 3 60   MODEL NM W=45U L=4U
Bsim M3b  1 22 3 60  MODEL NM W=45U L=4U
Bsim Mb   7 6 3 60   MODEL NM W=30U L=4U
Bsim MB9  3 21 60 60   MODEL NM W=120U L=8U
Volt Vb 6 0 DC=1.6 I=I_Vb
*
$***Bias Circuit 2*******************************
Bsim M3   11 11 60 60  MODEL NM W=30U L=8U
Bsim M4   13 11 12 1   MODEL P1 W=75U L=8U
Bsim M5   60 21 84 1   MODEL P1 W=75U L=8U
Bsim M7   13 13 60 60  MODEL NM W=40U L=15U
Bsim M8   5 13 60 60   MODEL NM W=80U L=15U
Bsim M9   84 13 60 60  MODEL NM W=40U L=15U
Bsim M10  16 16 84 1   MODEL P1 W=50U L=4U
Bsim M11  17 16 12 1   MODEL P1 W=50U L=4U
Bsim M12  16 17 60 60  MODEL NM W=20U L=5U
Bsim M13  17 17 60 60  MODEL NM W=20U L=5U
Curr IBIAS2 1 84 DC=25U
Curr IBIAS3 1 12 DC=20U
*
$***Inmax--Ipmax Circuit*************************
Bsim MB1  21 21 60 60  MODEL NM W=30.0U L=8U PUBLIC
Bsim MB2  80 27 60 60  MODEL NM W=30U L=8U PUBLIC
Bsim MB3  27 27 20 1 MODEL P1 W=75U L=8U
Bsim MB4  26 21 20 1 MODEL P1 W=75U L=8U
Bsim MB5  27 26 60 60  MODEL NM W=25U L=20U  PUBLIC
Bsim MB6  26 26 60 60  MODEL NM W=100U L=20U PUBLIC
Bsim MB7  81 8 1 1   MODEL P1 W=50U L=10U
Bsim MB8  8 8 1 1    MODEL P1 W=200U L=10U
Bsim MB10  21 80 81 1 MODEL P1 W=50U L=2U
Bsim MB11 80 80 8 1  MODEL P1 W=200U L=2U
Curr IBIAS1 1 20 DC=25U
```

```
***Load*********************************************
Cap Cl1   72 500 5P
Res Rout  72 500 100000
Volt Vref 500 0 DC 1.5

***Compensation Caps*****
Cap Cc1 72 201 2P
Cap Cc2 72 203 2P
Cap Cc3 79 202 2P
Cap Cc4 79 200 2P

Volt VDD 1 0  DC=3
Volt VSS 60 0 DC=0
```

B.20 Opamp 3a

Circuit shown in Figure 8.11

```
***Output Stage******************************************
*
***Common Source Amplifier*************************
Bsim MO1 72 208 1 1   MODEL P1 W=100U L=2U PUBLIC
Bsim MO1B 91 217 1 1  MODEL P1 W=100U L=2U PUBLIC
Bsim MO2 72 79 60 60  MODEL NM W=40U  L=2U PUBLIC
Bsim MO2B 91 92 50 50 MODEL NM W=40U L=2U PUBLIC
*
***Class AB Control*********************************
Bsim MO3  1 76 75 60     MODEL NM W=100U L=4U PUBLIC
Bsim MO4  202 171 75 60  MODEL NM W=100U L=4U PUBLIC
Bsim MO4B 203 171 75 60  MODEL NM W=100U L=4U PUBLIC
Bsim MO5  79 79 60 60    MODEL NM W=20U  L=2U PUBLIC
Bsim MO5B 92 92 50 50    MODEL NM W=20U L=2U PUBLIC
Bsim MO6  79 217 1 1     MODEL P1 W=50U  L=2U PUBLIC
Bsim MO6B 92 208 1 1     MODEL P1 W=50U L=2U PUBLIC
Bsim MO7  77 78 60 60    MODEL NM W=40U  L=4U PUBLIC
Bsim MO7B 78 78 60 60    MODEL NM W=40U  L=4U PUBLIC
Bsim MO8  76 208 1 1     MODEL P1 W=50U  L=2U PUBLIC
Bsim MO9  73 79 60 60    MODEL NM W=20U  L=2U PUBLIC
Bsim MO10 77 79 60 60    MODEL NM W=20U  L=2U PUBLIC
Bsim MO11 73 73 74 1     MODEL P1 W=50U  L=4U PUBLIC
Bsim MO12 60 73 171 1    MODEL P1 W=50U  L=4U PUBLIC
```

```
Bsim M013 78 77 76 1      MODEL P1 W=50U  L=4U PUBLIC
Bsim M014 77 77 74 1      MODEL P1 W=50U  L=4U PUBLIC
*
Volt Vc5 74 0 DC=2.5
Curr Imin  1 171 DC 5U
Curr IB 75 0 DC=30U
***IB1 and IB2 can be supplied by MC1 and MC2,
***but they are separately placed in this file.
Curr IB1 1 202  DC=10U
Curr IB2 1 203  DC=10U

***Cascode Gain Stage******************************
Bsim MC1 202 204  1 1    MODEL P1 W=100U L=8U PUBLIC
Bsim MC2 203 204 1 1     MODEL P1 W=100U L=8U PUBLIC
Bsim MC3 217 205 202 1   MODEL P1 W=550U L=4U PUBLIC
Bsim MC4 208 205 203 1   MODEL P1 W=550U L=4U PUBLIC
Bsim MC5 217 207 200 60 MODEL NM W=200U L=4U PUBLIC
Bsim MC6 208 207 201 60 MODEL NM W=200U L=4U PUBLIC
Bsim MC7 200 206 60 60  MODEL NM W=100U L=8U PUBLIC
Bsim MC8 201 206 60 60  MODEL NM W=100U L=8U PUBLIC

Volt VC1 204 0 DC 1.55
Volt VC2 205 0 DC 1.3
Volt VC3 207 0 DC 1.8
Volt V06 206 0 DC 1.35


*********Constant-gm Input Stage*****************
*
$***Input Differential Pairs*********************
Bsim M1  200 2 4 1    MODEL P1 W=75U L=8U PUBLIC
Bsim M1a 201 22 4 1   MODEL P1 W=75U L=8U PUBLIC
Bsim M2  202 2 5 60   MODEL NM W=30U L=8U PUBLIC
Bsim M2a 203 22 5 60  MODEL NM W=30U L=8U PUBLIC
*
***The circuit which keeps the Icascode constant**
Bsim ME1 202 209 1 1  MODEL P1 W=60U L=10U  PUBLIC
Bsim ME2 203 209 1 1  MODEL P1 W=60U L=10U PUBLIC
Bsim ME3 209 209 1 1  MODEL P1 W=60U L=10U PUBLIC
Bsim ME4 200 11 60 60 MODEL NM W=30U L=8U
Bsim ME5 201 11 60 60 MODEL NM W=30U L=8U
Bsim ME6 209 13 60 60 MODEL NM W=30U L=5U
*
$***Monitor Circuit 1*****************************
Bsim Mp   4 8 1 1   MODEL P1 W=400U L=4U
```

```
Bsim Mq    44 8 1 1  MODEL P1 W=200U L=4U
Bsim M3a  11 2 44 1 MODEL P1 W=38U L=8U
Bsim M3b  11 22 44 1 MODEL P1 W=38U L=8U
*
$***Bias Circuit 2*******************************
Bsim M3 11 11 60 60   MODEL NM W=30U L=8U  PUBLIC
Bsim M4 13 11 12 1    MODEL P1 W=75U L=8U PUBLIC
Bsim M5 60 21 84 1    MODEL P1 W=75U L=8U PUBLIC
Bsim M7 13 13 60 60   MODEL NM W=30U L=5U PUBLIC
Bsim M8 5 13 60 60    MODEL NM W=60U L=5U PUBLIC
Bsim M9 84 13 60 60   MODEL NM W=30U L=5U
Bsim M10 16 16 84 1   MODEL P1 W=100U L=4U
Bsim M11 17 16 12 1   MODEL P1 W=100U L=4U
Bsim M12 16 17 60 60  MODEL NM W=40U L=12U
Bsim M13 17 17 60 60  MODEL NM W=40U L=12U
Curr IBIAS2 1 84 DC=25U
Curr IBIAS3 1 12 DC=20U
*
$***Inmax--Ipmax Circuit*************************
Bsim MB1  21 21 60 60  MODEL NM W=30.0U L=8U PUBLIC
Bsim MB2  80 27 60 60  MODEL NM W=30U L=8U PUBLIC
Bsim MB3  27 27 20 1   MODEL P1 W=75U L=8U
Bsim MB4  26 21 20 1   MODEL P1 W=75U L=8U
Bsim MB5  27 26 60 60  MODEL NM W=25U L=20U  PUBLIC
Bsim MB6  26 26 60 60  MODEL NM W=100U L=20U PUBLIC
Bsim MB7  81 8 1 1     MODEL P1 W=50U L=4U
Bsim MB8  8 8 1 1      MODEL P1 W=200U L=4U
Bsim MB9  21 80 81 1   MODEL P1 W=50U L=2U
Bsim MB10 80 80 8 1    MODEL P1 W=200U L=2U
Curr IBIAS1 1 20 DC=25U

***Load********************************************
Cap Cl1 72 500 5P
*Res Rl1 72 500 100000
Cap Cl2 91 500 5P
*Res Rl2 91 500 100000
Volt Vref 500 0 DC 1.5

*Res Rf1 2 224 100000000
*Res Rf2 22 225 100000000
Res Rf1 2 72 10000
Res Rf2 22 91 10000
Res Rin1 2 222 10000
Res Rin2 22 223 10000
*Cap Cin1 2 222 100000000
```

```
*Cap Cin2 22 223 100000000
*Ind Lf1 224 72 100000000
*Ind Lf2 225 91 100000000

***Compensation Caps*****
Cap Cc1 72 201 2P
Cap Cc2 72 203 2P
Cap Cc3 79 202 2P
Cap Cc4 79 200 2P
Cap Cc11 91 200 2P
Cap Cc22 91 202 2P
Cap Cc33 92 203 2P
Cap Cc44 92 201 2P

***Power Supply***********
Volt VDD 1 0  DC=3
Volt VSS 60 0 DC=0
```

B.21 Opamp 3B

Circuit shown in Figure 8.14

```
***Output Stage*******************************************
*
***Common Source Amplifier************************
Bsim M01 72 208 1 1   MODEL P1 W=100U L=2U PUBLIC
Bsim M01B 91 217 1 1  MODEL P1 W=100U L=2U PUBLIC
Bsim M02 72 79 60 60  MODEL NM W=40U  L=2U PUBLIC
Bsim M02B 91 92 50 50 MODEL NM W=40U L=2U PUBLIC
*
***Class AB Control*********************************
Bsim M03  1 76 75 60      MODEL NM W=100U L=4U PUBLIC
Bsim M04  202 171 75 60   MODEL NM W=100U L=4U PUBLIC
Bsim M04B 203 171 75 60   MODEL NM W=100U L=4U PUBLIC
Bsim M05  79 79 60 60     MODEL NM W=20U  L=2U PUBLIC
Bsim M05B 92 92 50 50     MODEL NM W=20U  L=2U PUBLIC
Bsim M06  79 217 1 1      MODEL P1 W=50U  L=2U PUBLIC
Bsim M06B 92 208 1 1      MODEL P1 W=50U  L=2U PUBLIC
Bsim M07  77 78 60 60     MODEL NM W=40U  L=4U PUBLIC
Bsim M07B 78 78 60 60     MODEL NM W=40U  L=4U PUBLIC
Bsim M08  76 208 1 1      MODEL P1 W=50U  L=2U PUBLIC
Bsim M09  73 79 60 60     MODEL NM W=20U  L=2U PUBLIC
```

```
Bsim MO10 77 79 60 60     MODEL NM W=20U  L=2U PUBLIC
Bsim MO11 73 73 74 1      MODEL P1 W=50U  L=4U PUBLIC
Bsim MO12 60 73 171 1     MODEL P1 W=50U  L=4U PUBLIC
Bsim MO13 78 77 76 1      MODEL P1 W=50U  L=4U PUBLIC
Bsim MO14 77 77 74 1      MODEL P1 W=50U  L=4U PUBLIC
*
Volt Vc5 74 0 DC=2.5
Curr Imin  1 171 DC 5U
Curr IB 75 0 DC=30U
***IB1 and IB2 can be supplied by MC1 and MC2,
***but they are separately placed in this file.
Curr IB1 1 202  DC=10U
Curr IB2 1 203  DC=10U

***Cascode Gain Stage******************************
Bsim MC1 202 204  1 1   MODEL P1 W=100U L=8U PUBLIC
Bsim MC2 203 204 1 1    MODEL P1 W=100U L=8U PUBLIC
Bsim MC3 217 205 202 1  MODEL P1 W=550U L=4U PUBLIC
Bsim MC4 208 205 203 1  MODEL P1 W=550U L=4U PUBLIC
Bsim MC5 217 207 200 60 MODEL NM W=200U L=4U PUBLIC
Bsim MC6 208 207 201 60 MODEL NM W=200U L=4U PUBLIC
Bsim MC7 200 206 60 60  MODEL NM W=100U L=8U PUBLIC
Bsim MC8 201 206 60 60  MODEL NM W=100U L=8U PUBLIC

Volt VC1 204 0 DC 1.55
Volt VC2 205 0 DC 1.3
Volt VC3 207 0 DC 1.8
Volt V06 206 0 DC 1.35

***Input Stage**************************************
*
$***Input Differential Pairs**********************
Bsim M1  200 2 4 1    MODEL P1 W=75U L=8U PUBLIC
Bsim M1a 201 22 4 1   MODEL P1 W=75U L=8U PUBLIC
Bsim M2  202 2 5 60   MODEL NM W=30U L=8U PUBLIC
Bsim M2a 203 22 5 60  MODEL NM W=30U L=8U PUBLIC
*
***The circuit which keeps the Icascode constant**
Bsim ME1 202 209 1 1  MODEL P1 W=60U L=10U  PUBLIC
Bsim ME2 203 209 1 1  MODEL P1 W=60U L=10U PUBLIC
Bsim ME3 209 209 1 1  MODEL P1 W=60U L=10U PUBLIC
Bsim ME4 200 11 60 60 MODEL NM W=30U L=8U
Bsim ME5 201 11 60 60 MODEL NM W=30U L=8U
Bsim ME6 209 13 60 60 MODEL NM W=30U L=5U
*
```

```
$***Monitor Circuit 2****************************
Bsim Mp   4 7 1 1   MODEL P1 W=200U L=5U
Bsim Mq   11 7 1 1  MODEL P1 W=100U L=5U
Bsim Mr   7 7 1 1   MODEL P1 W=100U L=5U
Bsim M3a  1 2 3 60   MODEL NM W=45U L=4U
Bsim M3b  1 22 3 60  MODEL NM W=45U L=4U
Bsim Mb   7 6 3 60   MODEL NM W=30U L=4U
Bsim MB9  3 21 60 60   MODEL NM W=120U L=8U
Volt Vb 6 0 DC=1.6 I=I_Vb
*
$***Bias Circuit 2*******************************
Bsim M3   11 11 60 60  MODEL NM W=30U L=8U
Bsim M4   13 11 12 1   MODEL P1 W=75U L=8U
Bsim M5   60 21 84 1   MODEL P1 W=75U L=8U
Bsim M7   13 13 60 60  MODEL NM W=40U L=15U
Bsim M8   5 13 60 60   MODEL NM W=80U L=15U
Bsim M9   84 13 60 60  MODEL NM W=40U L=15U
Bsim M10  16 16 84 1   MODEL P1 W=50U L=4U
Bsim M11  17 16 12 1   MODEL P1 W=50U L=4U
Bsim M12  16 17 60 60  MODEL NM W=20U L=5U
Bsim M13  17 17 60 60  MODEL NM W=20U L=5U
Curr IBIAS2 1 84 DC=25U
Curr IBIAS3 1 12 DC=20U
*
$***Inmax--Ipmax Circuit*************************
Bsim MB1  21 21 60 60  MODEL NM W=30.0U L=8U PUBLIC
Bsim MB2  80 27 60 60  MODEL NM W=30U L=8U PUBLIC
Bsim MB3  27 27 20 1 MODEL P1 W=75U L=8U
Bsim MB4  26 21 20 1 MODEL P1 W=75U L=8U
Bsim MB5  27 26 60 60  MODEL NM W=25U L=20U  PUBLIC
Bsim MB6  26 26 60 60  MODEL NM W=100U L=20U PUBLIC
Bsim MB7  81 8 1 1   MODEL P1 W=50U L=10U
Bsim MB8  8 8 1 1    MODEL P1 W=200U L=10U
Bsim MB10  21 80 81 1 MODEL P1 W=50U L=2U
Bsim MB11 80 80 8 1  MODEL P1 W=200U L=2U
Curr IBIAS1 1 20 DC=25U

***Load*****************************************
Cap Cl1 72 500 5P
*Res Rl1 72 500 100000
Cap Cl2 91 500 5P
*Res Rl2 91 500 100000
Volt Vref 500 0 DC 1.5

*Res Rf1 2 224 100000000
```

```
*Res Rf2 22 225 100000000
Res Rf1 2 72 10000
Res Rf2 22 91 10000
Res Rin1 2 222 10000
Res Rin2 22 223 10000
*Cap Cin1 2 222 100000000
*Cap Cin2 22 223 100000000
*Ind Lf1 224 72 100000000
*Ind Lf2 225 91 100000000

***Compensation Caps*****
Cap Cc1 72 201 2P
Cap Cc2 72 203 2P
Cap Cc3 79 202 2P
Cap Cc4 79 200 2P
Cap Cc11 91 200 2P
Cap Cc22 91 202 2P
Cap Cc33 92 203 2P
Cap Cc44 92 201 2P


Volt VDD 1 0  DC=3
Volt VSS 60 0 DC=0
```

Appendix C

Measurement Techniques

In this Appendix, some of the measurement techniques used to evaluate the test chips are provided.

C.1 Input Stage Transconductance Measurements

To measure the transconductance of a transistor, whose source terminal is connected to a fixed voltage(as shown in Figure C.1(a)), one can simply increment the gate voltage and observe the change in the drain current. That is, g_m of the transistor at $V_g = V_{gO}$ can be found by sweeping V_g from $V_{gO} - \triangle V_g/2$ to $V_{gO} + \triangle V_g/2$ and measuring I_d at each V_g. Then,

$$g_m = \frac{\triangle I_d}{\triangle(V_g - V_s)}. \tag{C.1}$$

The procedure can be repeated for different values of V_{gO}, and one can obtain a plot of g_m as a function of V_g. However, we are interested in g_m of the input differential pairs whose source terminals are connected to current sources. Figure C.1(b) shows a similar situation where M_b is acting as a current source and g_m of M_a needs to be measured as a function of V_g. It seems that we could again sweep V_g and determine g_m from (C.1), but for the following reason, this scheme does not work. For V_g near V_{s2}, I_d is small because M_b is in the triode region. As V_g is increased, so does I_d and V_{gs}, and we will be able to find g_m. However, once V_g is sufficiently large such that M_b is in the saturation region, I_d becomes constant and so does V_{gs}, and the quantity given by (C.1) will be strongly affected by the output resistance of M_b. For instance, if the current source was ideal, I_d

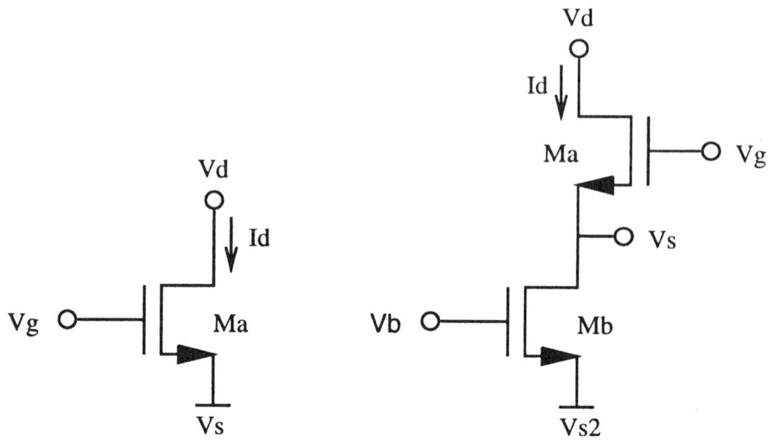

(a) (b)

Figure C.1: Transistors whose g_m are to be measured. (a) Source terminal is fixed. (b) Source terminal is dependent on V_g.

will be perfectly constant regardless of V_g, and $\triangle I_d$ and $\triangle V_{gs}$ will always be zero. Figure C.2 shows the simulated transconductance of M_a in Figure C.1(b); the curve labeled "direct" was obtained simply by asking APLAC to determine g_m of M_a while the curve labeled "indirect" was obtained by using (C.1). We can clearly see that as M_b becomes closer to being in the saturation region, g_m determined by (C.1) suddenly decreases.

In order to overcome the problem associated with the above scheme for measuring g_m of the input transistors, the following technique was used. We must use the circuit shown in Figure C.1(b) because that is the way the differential pairs are used; however, we make M_a detachable from the current source(drain of M_a must be open so that I_d can be measured.) To find g_m of M_a at $V_g = V_{gO}$, first, with M_a connected to M_b, set $V_g = V_{gO}$. Measure $I_d = I_d(V_{gO})$ and $V_s = V_s(V_{gO})$. Now, disconnect M_a from M_b and set $V_g = V_{gO}$ and $V_s = V_s(V_{gO})$ found above; at this point, I_d should also equal to $I_d(V_{gO})$ found above. Note that we now have the situation shown in Figure C.1(a), and thus (C.1) can be used to determine g_m at V_{gO}. The above technique was used to measure g_m of the circuit shown in Figure C.1(b) on a test chip; this is shown in Figure C.3 and is compared with the simulation result(the "direct" curve of Figure C.2.) Note that the results are close and enhance the validity of both the simulation and the measurement techniques.

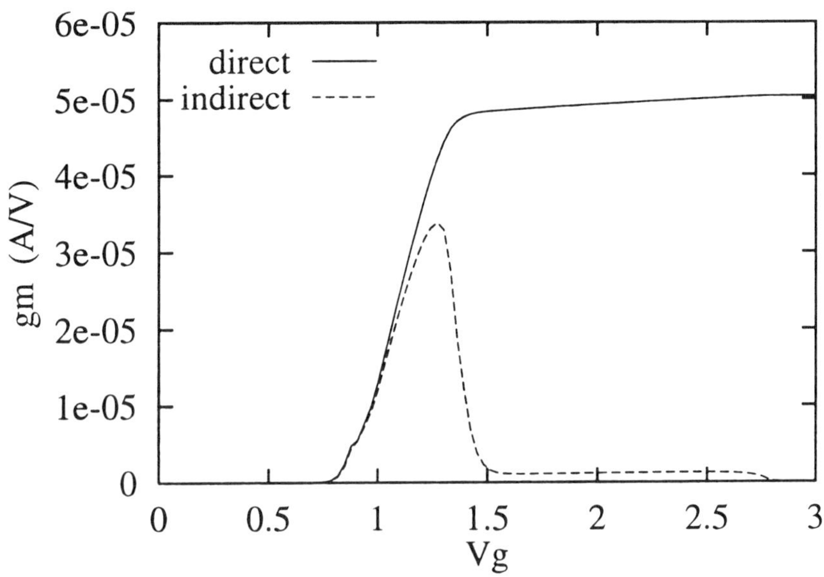

Figure C.2: Transconductance of M_a simulated using two different schemes.

C.2 Low Frequency Operational Amplifier Gain Measurements

The circuit shown in Figure C.4[63] was used to measure the low frequency(50Hz) open loop gain of the operational amplifiers. The opamp to be tested is connected in an inverting gain mode. The buffer consisting of CA3160 opamp in the feedback path avoids the loading of the opamp output node by R_2. The noise signal v_s is applied at the input of the inverting amplifier, and thus the output voltage is given by

$$v_o = -Av_{in} \tag{C.2}$$

$$= -\frac{R_2}{R_1}v_s \tag{C.3}$$

The opamp gain A can be simply determined by measuring v_{in} and v_o. These signals are again buffered by another CA3160 opamp; since we are concerned with the ratio of v_o to v_{in}, any nonideal effect introduced by

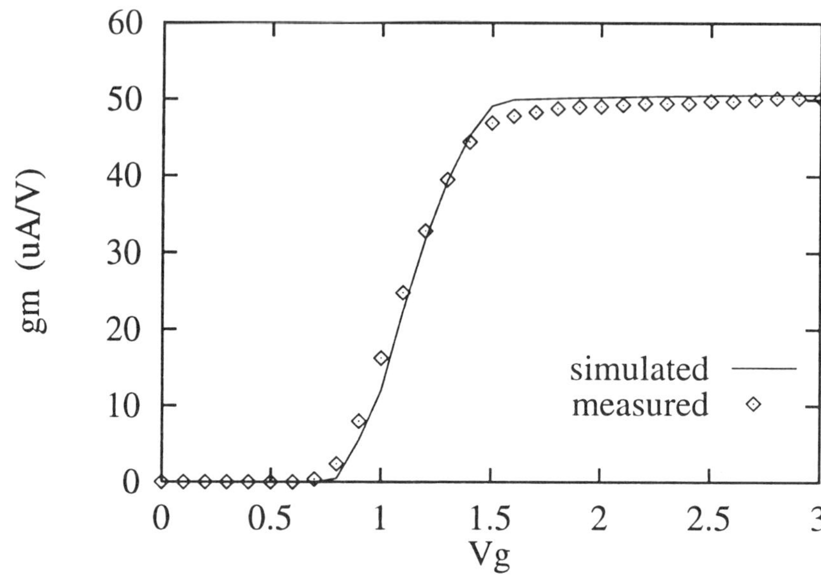

Figure C.3: Transconductance of M_a measured from the test chip.

CA3160 is cancelled. However, it was found from experiments, that the results are reliable only at low frequencies. Thus, this scheme was used only for the measurement of the low frequency gain of the opamps.

C.3 Unity Gain Frequency and Phase Margin Measurements

An accurate way to determine the unity gain frequency, f_U, and the phase margin, ϕ_M, of operational amplifiers is given next. Let A be the transfer function of the opamp and let us write it in the following form:

$$A = a + ib \tag{C.4}$$

where a and b are the real and the imaginary part, respectively, of the transfer function. Then the magnitude of A is given by

$$|A| = \sqrt{a^2 + b^2} \tag{C.5}$$

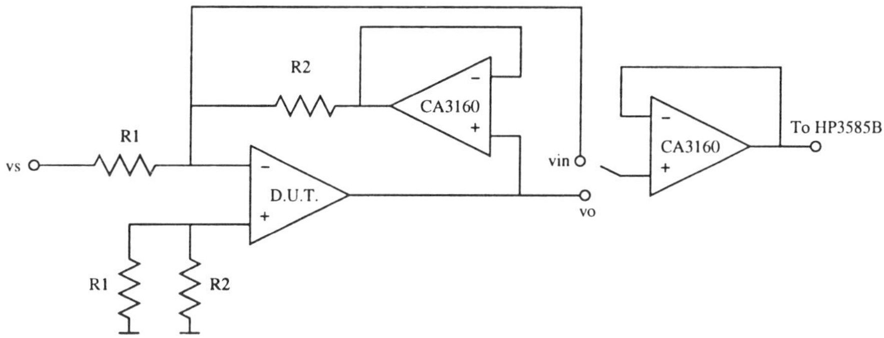

Figure C.4: Circuit used to measure the low frequency open loop gain of the opamps.

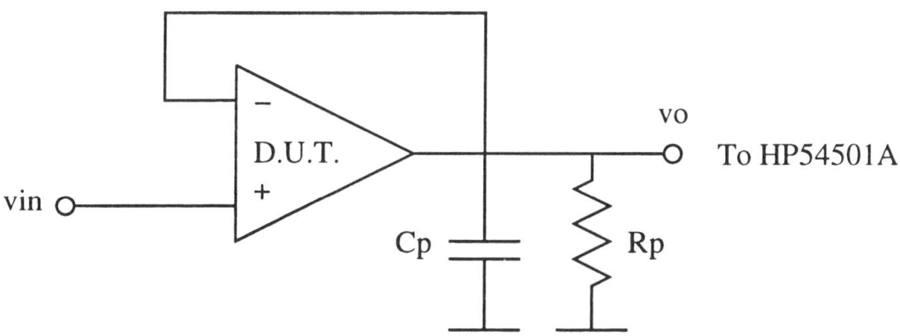

Figure C.5: Circuit used to measure f_U and ϕ_M of the opamps.

and the phase of A is given by

$$\phi_A = tan^{-1}(b/a). \qquad (C.6)$$

Now, consider a unity gain buffer connected opamp as shown in Figure C.5. The output node is loaded with R_p and C_p which are the resistance and capacitance, respectively, seen by the output node; C_p includes the pin capacitance of the output and the inverting input terminals of the opamp, the capacitances from the bread board and coaxial cable, and the input capacitance of the HP54501A, R_p is the input resistance of the HP54501A. C_p and R_p are approximately $20pF$ and $1M\Omega$. The transfer function of the buffer is given by

$$\frac{v_o}{v_{in}} = \frac{A}{1+A} \qquad (C.7)$$

and by substituting (C.4) into (C.7), we find the following.

$$|v_o/v_{in}|^2 = \frac{a^2+b^2}{1+2a+a^2+b^2} \tag{C.8}$$

$$\phi_{v_o/v_{in}} = tan^{-1}(\frac{b}{a+a^2+b^2}). \tag{C.9}$$

At f_U, which is the frequency of our interest, the opamp gain is one and (C.5) equals to one. Then substituting $a^2+b^2=1$ into (C.8) and (C.9), we get

$$|v_o/v_{in}|^2 = \frac{1}{2+2a} \tag{C.10}$$

$$\phi_{v_o/v_{in}} = tan^{-1}(\frac{\sqrt{1-a}}{\sqrt{1+a}}). \tag{C.11}$$

Note that (C.10) and (C.11) are both satisfied at f_U, and they are expressed using only one variable, a. Thus, f_U can be found by applying v_{in} with known frequency and checking if the above equations are satisfied. In order to do this, one can generate a table of $|v_o/v_{in}|$ and $\phi_{v_o/v_{in}}$ as a function of a, and search for an input signal frequency that results in the measured $|v_o/v_{in}|$ and $\phi_{v_o/v_{in}}$ to match one of the entries in the table(part of such a table is listed in Table C.2.) Once such a frequency is found, it is f_U, and a can be read off the table and ϕ_A and hence ϕ_M can be determined. As an example, procedure for determining f_U of opamp 1b with $V_{CM}=2.1V$ is given as follows. As listed in Table C.1, when the frequency of the input signal is $0.5MHz$, $|v_o/v_{in}|=0.67$ and $\phi_{v_o/v_{in}}=-36.9^O$. Now we look at Table C.2 and see that for $|v_o/v_{in}|=0.67$, $\phi_{v_o/v_{in}}$ should be 42°. This means that f_U is not exactly $0.5MHz$. The procedure is repeated at different frequencies until we have picked a correct frequency for the input signal satisfying both magnitude and phase, and in this case it is $0.53MHz$. Note that this is a time consuming technique, but since we do not have to make any assumptions regarding the transfer function of the opamp, the results are reliable.

Table C.1: Measurement results of opamp 1b as a function of the input signal frequency.

$f(MHz)$	$\lvert v_o/v_{in} \rvert$	$\phi_{v_o/v_{in}}(^O)$
0.50	0.67	-36.9
0.52	0.66	-38.4
0.53	0.65	-39.1

Table C.2: A table used to determine f_U and ϕ_M.

a	$\lvert v_o/v_{in} \rvert$	$\phi_{v_o/v_{in}}(^O)$	$\phi_M(^O)$
0.0400	0.6934	-43.8537	92.2926
0.0600	0.6868	-43.2800	93.4400
0.0800	0.6804	-42.7056	94.5888
0.1000	0.6742	-42.1303	95.7394
0.1200	0.6682	-41.5539	96.8923
0.1400	0.6623	-40.9760	98.0480
0.1600	0.6565	-40.3965	99.2071
0.1800	0.6509	-39.8150	100.3699
0.2000	0.6455	-39.2314	101.5371
0.2200	0.6402	-38.6454	102.7092
0.2400	0.6350	-38.0566	103.8867
0.2600	0.6299	-37.4649	105.0702
0.2800	0.6250	-36.8698	106.2604
0.3000	0.6202	-36.2711	107.4578

Bibliography

[1] B. Prince and R. H. Salters, "IC voltage dives," *IEEE Spectrum*, vol. 29, pp. 22 – 25, May 1992.

[2] R. H. Dennard, F. H. Gaensslen, H. N. Yu, V. L. Rideout, E. Bassous, and A. R. LeBlanc, "Design of ion–implanted MOSFET's with very small physical dimensions," *IEEE Journal of Solid - State Circuits*, vol. SC-9, pp. 256 – 268, October 1974.

[3] Y. A. El-Mansy, "On scaling MOS devices for VLSI," *IEEE Int. Conf. on Circuits and Computers*, vol. 1, pp. 457 – 460, 1980.

[4] J. E. Chung, M. C. Jeng, J. E. Moon, P. K. Ko, and C. Hu, "Performance and reliability design issue for deep-submicrometer MOSFET's," *IEEE Transactions on Electron Devices*, vol. 38, pp. 545 – 554, March 1991.

[5] T. H. Ning, P. W. Cook, R. H. Dennard, C. M. Osburn, S. E. Schuster, and H. N. Yu, "1 μm MOSFET VLSI technology: part IV–hot-electron design constraints," *IEEE Journal of Solid - State Circuits*, vol. SC-14, pp. 268– 275, April 1979.

[6] C. Hu, S. C. Tam, F. C. Hsu, P. K. Ko, T. Y. Chan, and K. W. Terrill, "Hot-electron-induced MOSFET degradation–model, monitor, and improvement," *IEEE Transactions on Electron Devices*, vol. ED-32, pp. 375 – 385, February 1985.

[7] E. A. Vittoz, "The design of high-performance analog circuits on digital CMOS chips," *IEEE Journal of Solid - State Circuits*, vol. SC-20, pp. 657 – 665, June 1985.

[8] E. Sano, T. Tsukahara, and A. Iwata, "Performance limits of mixed analog/digital circuits with scaled MOSFET's," *IEEE Journal of Solid - State Circuits*, vol. 23, pp. 942 – 949, August 1988.

[9] J. Y. Michel, "High-performance analog cells in mixed-signal VLSI: problems and practical solutions," *Analog Integrated Circuits and Signal Processing*, vol. 1, pp. 171 – 182, November 1991.

[10] S. Wong and C. A. T. Salama, "Impact of scaling on MOS analog perfermance," *IEEE Journal of Solid - State Circuits*, vol. SC-18, pp. 106 – 114, February 1983.

[11] T. Enomoto, T. Ishihara, M. Yasumoto, and T. Aizawa, "Design, fabrication, and performance of scaled analog IC's," *IEEE Journal of Solid - State Circuits*, vol. SC-18, pp. 395 – 402, August 1983.

[12] C. G. Sodini, S. S. Wong, and P. K. Ko, "A framework to evaluate technology and device design enhancements for MOS integrated circuits," *IEEE Journal of Solid - State Circuits*, vol. 24, pp. 118 – 127, February 1989.

[13] D. J. Allstot, R. W. Brodersen, and P. R. Gray, "MOS switched capacitor ladder filters," *IEEE Journal of Solid - State Circuits*, vol. SC-13, pp. 806 –814, December 1987.

[14] R. W. Brodersen, P. R. Gray, and D. A. Hodges, "MOS switched-capacitor filters," *Proceeding of the IEEE*, vol. 67, pp. 61 – 75, January 1979.

[15] K. Martin, "Improved circuits for the realization of switched-capacitor filters," *IEEE Transactions on Circuits and Systems*, vol. CAS-27, pp. 237 – 244, April 1980.

[16] T. C. Choi, R. T. Kaneshiro, R. W. Brodersen, P. R. Gray, W. B. Jett, and M. Wilcox, "High-frequency CMOS switched-capacitor filters for communications application," *IEEE Journal of Solid - State Circuits*, vol. SC-18, pp. 652 – 664, December 1983.

[17] R. Castello and P. R. Gray, "A high-performance micropower switched-capacitor filter," *IEEE Journal of Solid - State Circuits*, vol. SC-20, pp. 1122 – 1132, December 1985.

[18] Y. Tsividis, M. Banu, and J. Khoury, "Continuous-time MOSFET-C filters in VLSI," *IEEE Transactions on Circuits and Systems*, vol. CAS-33, pp. 125 – 139, February 1986.

[19] M. Ismail, S. V. Smith, and R. G. Beale, "A new MOSFET-C universal filter structure for VLSI," *IEEE Journal of Solid - State Circuits*, vol. 23, pp. 183 –194, February 1988.

[20] S. Sakurai, M. Ismail, J.-Y. Michel, E. Sanchez-Sinencio, and R. Brannen, "A MOSFET–C variable equalizer circuit with simple on–chip automatic tuning," *IEEE Journal of Solid - State Circuits*, vol. 2, pp. 927 - 934, June 1992.

[21] F. Krummenacher and N. Joehl, "A 4-MHz CMOS continuous-time filter with on-chip automatic tuning," *IEEE Journal of Solid - State Circuits*, vol. 23, pp. 750 - 758, June 1988.

[22] E. Sanchez-Sinencio, R. L. Geiger, and H. Nevarez-Lozano, "Generation of continuous-time two integrator loop OTA filter structures," *IEEE Transactions on Circuits and Systems*, vol. 35, pp. 936 - 946, August 1988.

[23] B. Nauta, *Analog CMOS filters for very-high frequencies*. PhD thesis, University of Twente, 1991.

[24] K. H. Loh, D. L. Hiser, W. J. Adams, and R. L. Geiger, "A versatile digitally controlled continuous-time filter structure with wide-range fine resolution capability," *IEEE Trans. CAS-II*, vol. 39, pp. 265 - 276, May 1992.

[25] A. Nedungadi and T. R. Viswanathan, "Design of linear CMOS transconductance elements," *IEEE Transactions on Circuits and Systems*, vol. CAS-31, pp. 891 - 894, October 1984.

[26] R. R. Torrance, T. R. Viswanathan, and J. H. Hanson, "CMOS voltage to current transducers," *IEEE Transactions on Circuits and Systems*, vol. CAS-32, pp. 1097 - 1985, November 1985.

[27] E. Seevinck and R. F. Wassenaar, "A versatile CMOS linear transconductor/square-law function circuits," *IEEE Journal of Solid - State Circuits*, vol. SC-32, pp. 366 - 377, June 1987.

[28] P. M. V. Peteghem, H. M. Fossati, G. L. Rice, and S. Y. Lee, "Design of a very linear CMOS transconductance input stage for continuous-time filters," *IEEE Journal of Solid - State Circuits*, vol. 25, pp. 497 - 501, April 1990.

[29] Z. Wang and W. Guggenbuhl, "A voltage-controllable linear MOS transconductor using bias offset technique," *IEEE Journal of Solid - State Circuits*, vol. 25, pp. 315 - 317, February 1990.

[30] G. Wilson and P. K. Chan, "Low-distortion CMOS transconductor," *Electron. Lett.*, vol. 26, pp. 720 - 722, 1990.

[31] S. T. Dupuie and M. Ismail, "High frequency CMOS transconductors," in *Analogue IC Design: The Current-Mode Approach* (Toumazou, Lidgey, and Haigh, eds.), ch. 5, Peter Peregrinus Ltd, London, 1990.

[32] Z. Wang, "Making CMOS OTA a linear transconductor," *Electron. Lett.*, vol. 26, pp. 1448 - 1449, August 1990.

[33] Z. Wang, "Novel linearisation technique for implementing large-signal MOS tunable transconductor," *Electron. Lett.*, vol. 26, pp. 138 - 139, January 1990.

[34] S. Szczepanski, R. Schaumann, and P. Wu, "Linear transconductor based on crosscoupled CMOS pairs," *Electron. Lett.*, vol. 27, pp. 783 - 784, April 1991.

[35] P. Wu and R. Schaumann, "Tunable operational transconductance amplifier with extremely high linearity over very large input range," *Electron. Lett.*, vol. 27, pp. 1254 - 1255, 1991.

[36] E. J. van der Zwan, E. A. M. Klumperink, and E. Seevinck, "A CMOS OTA of HF filters with programmable transfer function," *IEEE Journal of Solid - State Circuits*, vol. 26, p. 1720, November 1991.

[37] R. Hogervorst, R. J. Wiegerink, P. A. de Jong, J. Fonderie, R. F. Wassenaar, and J. H. Huijsing, "CMOS low-voltage operational amplifiers with constant-g_m rail-to-rail input stage," *Proceedings of the IEEE International Symposium on Circuits and Systems*, pp. 2876 - 2879, 1992.

[38] J. H. Botma, R. F. Wassenaar, and R. J. Wiegerink, "A low-voltage CMOS op amp with a rail-to-rail constant-g_m input stage and a class AB rail-to-rail output stage," *Proceedings of the IEEE International Symposium on Circuits and Systems*, pp. 1314 - 1317, 1993.

[39] Helsinki University of Technology, Circuit Theory Laboratory and Nokia Research Center, *APLAC - An Object - Oriented Analog Circuit Simulator and Design Tool, 6.0 User's Manual and Reference Manual*, October 1991.

[40] P. R. Gray and R. G. Meyer, "MOS operational amplifier design–a tutorial overview," *IEEE Journal of Solid - State Circuits*, vol. SC-17, pp. 969 - 982, December 1982.

[41] J. E. Solomon, "The monolithic op amp: a tutorial study," *IEEE Journal of Solid - State Circuits*, vol. SC-9, pp. 314 – 332, December 1974.

[42] J. H. Huijsing and D. Linebarger, "Low-voltage operational amplifier with rail-to-rail input and output range," *IEEE Journal of Solid - State Circuits*, vol. SC-20, pp. 1144 – 1150, December 1985.

[43] J. Fonderie, M. M. Maris, and E. J. Schnitger, "1-V operational amplifier with rail-to-rail input and output ranges," *IEEE Journal of Solid - State Circuits*, vol. 24, pp. 1551 – 1559, December 1989.

[44] M. J. Fonderie and J. H. Huijsing, *Design of low-voltage bipolar operational amplifiers.* Kluwer Academic Publishers, 1993.

[45] J. A. Fisher and R. Koch, "A highly linear CMOS buffer amplifier," *IEEE Journal of Solid - State Circuits*, vol. SC-22, pp. 330 – 334, June 1987.

[46] M. Steyaert and W. Sansen, "A high-dynamic-range CMOS op amp with low-distortion output structure," *IEEE Journal of Solid - State Circuits*, vol. SC-22, pp. 1204 – 1207, December 1987.

[47] J. N. Babanezhad, "A rail-to-rail CMOS op amp," *IEEE Journal of Solid - State Circuits*, vol. 23, pp. 1414 – 1417, December 1988.

[48] T. S. Fiez, H. C. Yang, J. J. Yand, C. Yu, and D. J. Allstot, "A family of high-swing CMOS operational amplifiers," *IEEE Journal of Solid - State Circuits*, vol. 24, pp. 1683 – 1687, December 1989.

[49] M. D. Pardoen and M. G. Degrauwe, "A rail-to-rail input/output CMOS power amplifier," *IEEE Journal of Solid - State Circuits*, vol. 25, pp. 501 – 504, April 1990.

[50] R. Hogervorst, R. J. Wiegerink, P. A. D. Jong, J. Fonderie, R. F. Wassenaar, and J. H. Huijsing, "CMOS low-voltage operational amplifiers with constant-g_m rail-to-rail input stage," *Analog Integrated Circuits and Signal Processing*, vol. 5, pp. 135 – 146, March 1994.

[51] J. Huijsing, R. Hogervorst, J. Fonderie, B. van den Dool, K.-J. de Langen, and G. Groenevold, "Low-voltage signal processing," in *Analog VLSI: Signal and Information Processing* (Ismail and Fiez, eds.), ch. 4, McGraw Hill, 1994.

[52] M. J. M. Pelgrom, A. C. J. Duinmaiger, and A. P. G. Welbers, "Matching properties of MOS transistors for precision analog design," *IEEE Journal of Solid - State Circuits*, vol. SC - 24, pp. 1433 – 1439, October 1989.

[53] C. Michael and M. Ismail, "Statistical modeling of device mismatch for analog MOS integrated circuits," *IEEE Journal of Solid - State Circuits*, vol. SC - 27, pp. 154 – 166, February 1992.

[54] S. Sakurai and M. Ismail, "Constant transconductance bias circuit and method." U. S. Patent pending, filed August 25, 1993.

[55] A. Andreou and K. Boahen, "Neural information processing II," in *Analog VLSI: Signal and Information Processing* (Ismail and Fiez, eds.), ch. 8, McGraw Hill, 1994.

[56] P. Antognetti and G. Massobrio, *Semiconductor device modeling with SPICE*, ch. 4, pp. 163 – 164. McGraw Hill, 1988.

[57] D. B. Ribner and M. A. Copeland, "Design technique for cascoded CMOS op amp with improved PSRR and common-mode input range," *IEEE Journal of Solid - State Circuits*, vol. SC-19, pp. 919 – 925, December 1984.

[58] M. Banu, J. M. Khoury, and Y. Tsividis, "Fully differetial operational amplifiers with accurate output balancing," *IEEE Journal of Solid - State Circuits*, vol. 23, pp. 1410 – 1414, December 1988.

[59] S. M. Mallya and J. H. Nenin, "Design procedure for a fully differential folded-cascode CMOS operational amplifier," *IEEE Journal of Solid - State Circuits*, vol. 24, pp. 1737 – 1740, December 1989.

[60] C. Y. Wu, P. H. Lu, C. H. Lee, and M. K. Tsai, "New fully differentail HF CMOS op amp's with efficient common mode feedback," *Proceedings of the IEEE International Symposium on Circuits and Systems*, pp. 2076 – 2079, 1989.

[61] P. E. Allen and D. R. Holberg, *CMOS analog circuit design*, ch. 8. Holt, Rinehart and Winston, inc., 1987.

[62] A. L. Coban and P. E. Allen, "A 1.75V rail-to-rail CMOS op amp," *Proceedings of the IEEE International Symposium on Circuits and Systems*, pp. 497 – 500, 1994.

[63] W. M. C. Sansen, M. Steyaert, and P. J. V. Vandeloo, "Measurement of operational amplifier characteristics in the frequency domain," *IEEE Transactions on instrumentation and measurement*, vol. IM - 34, pp. 56 – 64, March 1985.

Index